DEADLY GLOW

THE RADIUM DIAL WORKER TRAGEDY

ROSS MULLNER, PhD, MPH

American Public Health Association
800 I Street, NW
Washington, DC 20001-3710

Mohammad N. Akhter, MD, MPH
Executive Vice President

© 1999 by the American Public Health Association

All rights reserved. This book is protected by copyright. No part of this book may be reproduced in any form or by any means, including photocopying, or utilized by any information storage and retrival system without prior written permission of the publisher.

2M   10/99
Library of Congress Catalog Card Number 99-073688

ISBN: 0-87553-245-4

Printed and bound in the United States of America.
Book Design and Typesetting: Jean-Marie Navetta
Production Support: Susan Westrate
Cover Design: Steve Trapero, Egad Dzyn
Set in: Adobe Garamond
Printing and Binding: Automated Graphics Systems, White Plains, Maryland

Table of Contents

Acknowledgments .... *v*
Foreword .... *vii*
Photo Credits .... *ix*
About the Author .... *xii*

   Introduction 1
1 Dawn of a Miracle 7
2 The First Nuclear Industry 15
3 Radium Medicine 31
4 Mysterious Deaths 41
5 Medical Detectives and Social Activists 55
6 In Search of Justice 75
7 The Ottawa Society of the Living Dead 91
8 The National Radium Scandal 109
9 Safety Standards and the Atomic Bomb 119
10 Under Radioactive Clouds 129
11 Conclusion 139

Appendix 145
Notes 153
Index 171

Acknowledgments

This book would not have been possible without the extremely generous assistance, advice, and encouragement of many people.

Specifically, I would like to thank John Rundo for writing the foreword; Albert Keane and LeAnna Westfall of the Argonne National Laboratory; the late Robley D. Evans, Professor Emeritus at the Massachusetts Institute of Technology; the late Glenn T. Seaborg of the University of California, Berkeley; Roger Cloutier and Paul Frame of the Oak Ridge Associated Universities; Lawrence Badash of the University of California, Santa Barbara; Edward R. Landa of the U.S. Geological Survey; Robert S. Harding of the Smithsonian Institution; George D. Tselos and Doug Tarr of the Edison National Historic Site; Barbara Irwin and Lois R. Densky-Wolff of the University of Medicine and Dentistry of New Jersey; Christopher H. Martland, Knoxville, Tennessee; Doctors Robert Bonda, Hyman W. Fisher, and William D. Sharp, formerly with the New Jersey Radium Research Project; Leonard J. Box, T & E Industries, Inc., Orange, New Jersey; George Kruto and Marguerite Fallucco of the American Medical Association; Dade W. Moeller of the Harvard School of Public Health; Janis Kiesig, the late Marie Rossiter, Edith Schomas, and Katie Troccoli, Ottawa, Illinois; Gertrude A. Grossman and Leonard Grossman, Chicago; Gary Albrecht and David Derenzo of the University of Illinois at Chicago; Janet Cappelli, John Prince, Matt Mankowski, and Rebecca Thomas of the various regional offices of the U.S. Environmental Protection Agency; and James Yusko of the Pennsylvania Department of Environmental Resources.

I am indebted to Alice M. Horowitz and Barbara Schaaf, who edited the manuscript and provided encouragement and enthusiasm. Finally, I would like to thank my wife, Linda, and my sons, Erik and Jason, for putting up with me while I finished the book.

RM

"Poisoned! as they chatted merrily at their work. *Painting the luminous numbers on watches, the radium accumulated in their bodies, and without warning began to bombard and destroy teeth, jaws, and finger bones, marking fifty young factory girls for painful, lingering, but inevitable death." (From* American Weekly, *Sunday newspaper insert, February 28, 1926, p. 11. Hearst Corporation.)*

Foreword

Deadly Glow is the story of an important tragedy. It is a story of the world's first widely known atomic victims—the radium dial workers. A generation before the first atomic bomb fell on Japan, the dial workers suffered and died from chronic radiation sickness. From their experience, the harmful effects of radium deposited in the body became known. The history of this tragedy needs to be told, and Dr. Mullner has delved into it with dogged determination. He has uncovered innumerable interesting and intriguing relevant facts, all of which are described in an easy-to-read style with little high-tech jargon.

How did this tragedy arise? The dial workers were mainly young women who were aged between as young as 11(!) and over 45 at first employment. In various plants in New Jersey, Connecticut, Illinois, and other states, these workers applied radium-activated luminous paint to watch and clock dials. In order to paint neat figures, the women twirled the brush between their lips, thus eating small amounts of the paint, and therefore, also of the radium. The latter turned out to accumulate in the skeleton because of its chemical similarity to calcium. The radiation emitted by the radium had disastrous effects initially on the jaw, but later also on other bones and tissues within the body.

Some good did come out of the tragedy. In 1934, Robley D. Evans, who became professor of physics at the Massachusetts Institute of Technology and director of the Radioactivity Center, was the first of a select band of scientists who made it their life's work to study radium poisoning in all of its terrible aspects. The timing was appropriate, because just a few years later, the phenomenon of nuclear fission in uranium was discovered in Germany and the possibility of atomic weapons was conceived and brought to fruition in the United States. This meant that huge amounts of radioactive materials were to be handled, and an understanding of the metabolism and effects of such materials became urgent and vital. The scientists from the Radioactivity Center provided much-needed information on these topics. There were two other centers where studies of the late effects of internally deposited radium were made: New Jersey and Argonne National Laboratory.

Evans retired in 1969, and the U.S. Atomic Energy Commission, following a suggestion made by him, proposed that all studies of radium (and other similar materials, notably plutonium) be consolidated at Argonne. This suggestion was accepted and, in 1969, the Center for Human Radiobiology was created as a section in the Radiological

Physics Division. I was privileged to become the center's head in 1980, a position I held on and off until I retired in 1991. The center was finally terminated in 1992. From 1969 to its end, scientists in the center were responsible for the publication in the scientific literature of more than 300 papers. We can now add the present book to this imposing list.

Who should read this book? In a word—well, two—almost everybody. It tells a fascinating tale.

John Rundo, D.Sc.
Former Section Head
Center for Human Radiobiology
Argonne National Laboratory

Photo Credits

Introduction

Figure 1a. A dial painter (front view) suffering from a radium-induced sarcoma of the chin.
(Author's collection)
Figure 1b. A dial painter (side view) suffering from a radium-induced sarcoma of the chin.
(Author's collection)
Figure 2. A dial painter with a radium-induced cancer of the knee
(Reproduced with Permission from Lippincott, Williams and Wilkins)

1. Dawn of A Miracle

Figure 3. Marie and Pierre Curie.
(William J. Hammer Collection, Archives Center, National Museum of American History, Smithsonian Institution)
Figure 4. Newspaper illustration entitled "Hope" from the Page Los Angeles Examiner, 1914.
(Reproduced from Argonne National Laboratory)

2. The First Nuclear Industry

Figure 5. William J. Hammer.
(William J. Hammer Collection, Archives Center, National Museum of American History, Smithsonian Institution)
Figure 6. Thomas Edison at his laboratory in West Orange, New Jersey.
(Edison National Historic Site, National Park Service, U.S. Department of the Interior, West Orange, NJ)
Figure 7. Area of major carnotite ore deposits in Colorado and Utah.
(Author's collection)
Figure 8. Stephen T. Lockwood.
(Buffalo Museum of Science, Buffalo, NY)
Figure 9. Joseph M. Flannery (fourth from left), James J. Flannery (fifth), and Henry Ford (seventh).
(Henry Ford Museum and Greefield Village Research Center, Dearborn, MI)
Figure 10. A Colorado radium mine.
(The Denver Public Library, Western History Collection)
Figure 11. Standard Chemical Company's radium refinery at Canonsburg, Pennsylvania.
(Reproduced from Argonne National Laboratory)
Figure 12. Chemists at the Standard Chemical Company, Pitsburgh.
(Reproduced from Argonne National Laboratory)
Figure 13. Marie Curie and President Harding outside the White House.
(National Archives, Washington, DC)

3. Radium Medicine

Figure 14. *A physician treating a lupus patient ith radium at the St. Louis Hospital in Paris.*
(Hisotry of Medicine Division, National Library of Medicine, Bethesda, MD)

Figure 15. *Revigator water jar (missing lid), ca. 1920.*
(Oak Ridge Associated Universities)

Figure 16. *Lifetime Radium Vitalizer, ca. 1926.*
(Oak Ridge Associated Universities)

Figure 17. *Radium Emanator, ca. 1925-1935.*
(Oak Ridge Associated Universities)

Figure 18. *An advertisement from the journal Radium, 1914.*
(Reproduced from Argonne National Laboratory)

4. Mysterious Deaths

Figure 19. *Dr. Sabin Arnold von Sochocky.*
(Reproduced from Argonne National Laboratory)

Figure 20. *Advertisement for Ingersoll Radiolite watches from the Saturday Evening Post, 1917.*
(Author's collection)

Figure 21. *Dial painter working at the U.S. Radium Corporation, ca. 1924.*
(Reproduced from Argonne National Laboratory)

Figure 22. *Dr. Theodor Blum.*
(Author's collection)

5. Medical Detectives and Social Activists

Figure 23. *Arthur Roeder.*
(Courtesy of Hagley Museum and Library, Wilmington, DE)

Figure 24. *Dr. Cecil K. Drinker.*
(Courtesy of Harvard School of Public Health and Dade W. Moeller)

Figure 25. *Florence Kelley.*
(Jane Addams Memorial Collection, Special Collections, University Library, University of Illinois at Chicago)

Figure 26. *Dr. Alice Hamilton.*
(Jane Addams Memorial Collection, Special Collections, University Library, University of Illinois at Chicago)

Figure 27. *Frederick L. Hoffman.*
(Courtesy of the Babson College Archives, Babson Park, MA)

Figure 28. *Dr. Martland (left) modeling a skull to help identify a homicide victim. Watching are Dr. Maurice Posers, Deputy ommissioner of the Royal Canadian mounted Police and, in the rear, Michael Frnzi, [Newark] City Hospital laboratory technician, February 1938.*
(University of Medicine and Dentistry of New Jersey Libraries, Special Collections, Harrison S. Martland Papers)

Figure 29. *Frederick B. Flinn.*
(Columbia University Archives and Columbiana Library)

Figure 30. *Dr. Martland at his desk with his "List of the Doomed," Life magazine, December 1951.*
(Bernard Hoffman, Life Magazine, Copyright Time, Inc.)

6. In Search of Justice

Figure 31. *Grace Fryer.*
(Reproduced from Argonne National Laboratory)

Figure 32. The five doomed women. Left to right: Quinta McDonald, Enda Hussman, Albina Larice, atherine Schaub, and Grace Fryer.
   (Reproduced from Argonne National Laboratory)
Figure 33. An editorial cartoon from the New York World criticizing the legal delays endured by the five doomed women, May 14, 1928.
   (Author's collection)

7. The Ottawa Society of the Living Dead

Figure 34. Young dial painters at the Peru Radium Dial Company, 1922.
   (Reproduced from Argonne National Laboratory)
Figure 35. Dial painters at the Ottawa Radium Dial Company, 1924.
   (Reproduced from Argonne National Laboratory)
Figure 36. Facsimile of the "Statement of the Radium Dial Company."
   (Reproduced with Permission of the Ottawa Daily Times)
Figure 37. Mary Ellen Cruse.
   (Reproduced with Permission of the Chicago Sun-Times)
Figure 38. Margaret Looney (left).
   (Author's collection)
Figure 39. Mrs. Donohue (center) and other members of the "Living Dead" signing a petition for Leonard Grossman to take their case, 1937.
   (Reproduced with Permission of the Chicago Sun-Times)
Figure 40. Mrs. Catherine Wolfe Donohue.
   (Reproduced with Permission of the Chicago Sun-Times)
Figure 41. The second day of the hearing held at the Donohue home. Mrs. Purcell can be seen behind Mr. Grossman.
   (Reproduced with Permission of the Chicago Sun-Times)

8. The National Radium Scandal

Figure 42. William J.A. Bailey.
   (Author's collection)
Figure 43. A bottle of Radithor.
   (Reproduced from Argonne National Laboratory)
Figure 44. Eben M. Byers.
   (Author's collection)

9. Safety Standards and the Atomic Bomb

Figure 45. Robley D. Evans and Nancy Caldwell demonstrating how a breath sample is taken to determine the amount of radium in the body, MIT, late 1950s.
   (MIT Museum, Cambridge, MA)
Figure 46. Glenn T. Seaborg.
   (Courtesy of Lawrence Berkeley National Laboratory and Glenn T. Seaborg)

10. Under Radioactive Clouds

Figure 47. The first atomic test in Nevada.
   (U.S. Army Photo, Library of Congress)
Figure 48. A 1950s postcard advertising the "Up and Atom" City of Las Vegas.
   (Lake County [IL] Museum, Curt Teich Postcard Archives)

About the Author

Ross Mullner, PhD, MPH, is an Associate Professor of Health Policy and Administration at the School of Public Health, University of Illinois at Chicago. He received a doctorate and two masters degrees from the University of Illinois. Dr. Mullner has published over eighty articles on various medical and public health topics. He formerly was a research fellow at Argonne National Laboratory, where he studied the radium dial workers. His interests include health services research, health care marketing, and the history of public health. He is currently writing a history of infectious diseases in Illinois.

Introduction

The radium dial painters' agony alerted the world to the hazards of internal radiation. ... They were the first victims of the wanton use of radioactive substances to alert the public to this new danger on the frontiers of science ... .[1]
Tony Bale, 1987

In the early 1920s a group of young women slowly and mysteriously began dying. The dying women seemed to have little in common, except that they all had previously worked as dial painters at a radium application plant in Orange, New Jersey. At the plant, the women painted the numerals on instrument and watch dials. The job seemed ideal. It paid well, depending upon the number of dials painted. And working with the new glowing radium paint was considered artistic, high-tech, and even glamorous.

Most of the women worked at the radium plant during World War I. The war created an enormous military demand for many types of radium-luminous devices. The nation's armed forces desperately needed radium dials for instruments aboard airplanes, submarines, and warships, and soldiers needed watches with glowing dials for night fighting.

Several years after leaving the plant, the former dial painters began developing a variety of mysterious medical problems. The women experienced abnormal blood changes, and they became severely anemic. They suffered from intense arthritic-like pains, particularly in the joints, which spread throughout their bodies. Some of the women suffered from spontaneous bone fractures of the arms and legs. A few of the former workers even became lame when their legs strangely began to shorten.

The most common symptoms they experienced, however, were horrible teeth and jaw problems. Typically, their teeth would ache constantly. And when a tooth was extracted, the socket would continue to bleed and not heal. Instead, it would slowly and painfully ulcerate. Eventually, the ulcer would spread and progressively worsen, leading to jaw necrosis, with parts of the women's jaws rotting away and disintegrating. Many times the necrosis would be so widespread that large sections of their jaws would have to be removed, in some cases leaving them horribly disfigured (Figures 1a and 1b).[2]

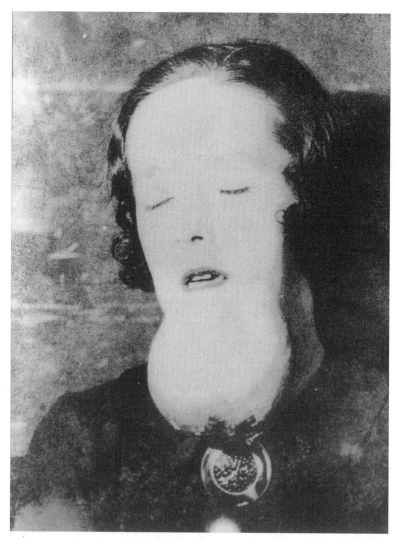

Figure 1a. *A dial painter (front view) suffering from a radium-induced sarcoma of the chin.*

Dentists who treated the suffering women were baffled by their mysterious disease. They quickly found, however, that they could do little to help them. In fact, they discovered that the more they attempted to treat the unknown medical condition, the faster the women's teeth and jaws deteriorated. Frustrated by their failure, several dentists went so far as to refuse to provide care to any women who previously worked as radium dial painters.

Physicians were equally baffled. The women's affliction did not seem to be like any other known medical condition. Their doctors mistakenly diagnosed them as suffering from various types of ulcers, a rare type of heart disease, and even the venereal disease, syphilis.

Introduction

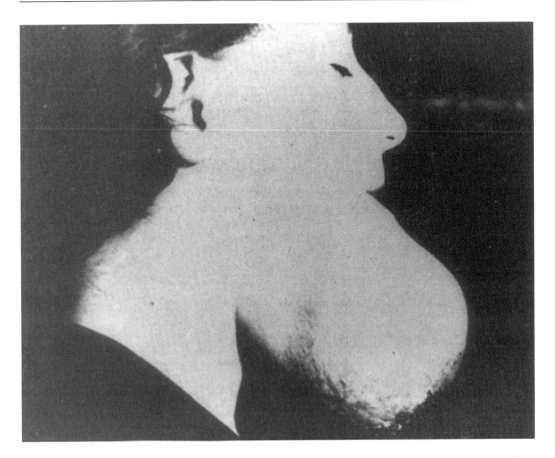

Figure 1b. *A dial painter (side view) suffering from a radium-induced sarcoma of the chin.*

One physician thought the strange disease might be occupational in origin. He diagnosed a former dial painter as suffering from white phosphorus poisoning, or "phossy jaw," an occupational disease known to result in the destruction of the teeth and jaws.[3] However, this disease was quickly ruled out when an analysis of the paint revealed it contained none of the substance.

Because they did not know the medical condition they were dealing with, the women's physicians treated the symptoms as best they could. The dial painters underwent a wide variety of medical tests, many surgical operations, and innumerable blood transfusions. But, nothing could stop the terrible ravages of the baffling disease.

When a few of the dying women went back to the radium plant to request that their former employer help them pay their staggering medical bills, the company steadfastly refused. The firm's managers said they were not at fault, denying any responsibility whatsoever for the women's health problems. Despite the overwhelming evidence to the contrary, the company would maintain this position for years. Not even the deaths from the disease of several of the firm's chemists and senior managers, including the corporation's co-founder, would alter the company's stand.

The president of the radium corporation vehemently argued that the women were not entitled to any compensation; they were not current workers, but former employees who had not worked for the company in many years. He indicated that the paint used at the factory had never contained white phosphorus or any other known toxic substances. And he summarily dismissed radium as causing the women's health problems.

He argued that the company's radium paint contained such an infinitesimal amount of the radioactive element that it could not possibly have caused any harm. Indeed, he retorted, many widely used medicines, tonics, and consumer products contained greater amounts of radium, and they had not caused any health problems.[4]

Several of the women applied for workers' compensation, only to find that they were ineligible. Under New Jersey law, the women had to file claims for their injuries either while they were still employed, or within two years after leaving the radium company. Yet many of the women had not suffered the first symptoms of the mysterious malady until several, in most cases four or more, years after their employment. Further, the state's inadequate workers' compensation law only covered a limited number of specific occupational diseases, which did not include their nameless illness.[5]

To receive compensation, the women's only resort was to sue the U.S. Radium Corporation for health damages. Filing a lawsuit against the large, powerful company, however, seemed hopeless. There were few legal precedents to point the way; and the radium corporation appeared to have a perfect legal defense. The radium corporation also could delay and drag out the legal proceedings for years. And even if their suits were successful, the women might not live long enough to receive any settlement.[6]

Despite their seemingly hopeless situation, a small number of women managed to file lawsuits against the corporation. Several of the suits, after extensive delays, would result in a number of sensational trials.

When Marie Curie learned of the radium dial painters' devastating illnesses, she bleakly stated that there was "practically no hope of saving the women." However, she did strongly reprimand the radium company, saying that she was surprised a plant with such bad hygienic conditions would exist in the United States.[7]

Florence Kelley, an influential social activist and former Chicago Hull House member who attempted to aid the radium dial painters, called the young women's horrible and lingering deaths nothing less than "cold-blooded murder in industry."[8]

Dr. Harrison S. Martland, a prominent New Jersey medical examiner and pioneering forensic pathologist who would study the dial painters for decades, found that with terrible, almost mathematical regularity, the women would die from the effects of the radium.[9]

Solving the intriguing mystery of the radium dial painters' baffling disease would be a complex and difficult task requiring brilliant detective work on the part of several investigators. The investigators would have the difficult task of confronting and challenging conventional wisdom and strongly held, but erroneous, beliefs concerning radium's safety and long-term health effects. They would have no real power or author-

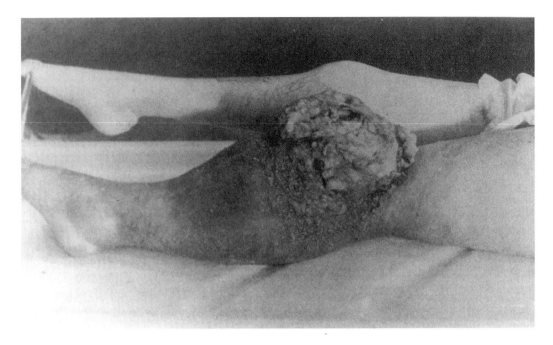

Figure 2. A dial painter with a radium-induced cancer of the knee.

ity to force the radium company to answer critical questions or to provide vital evidence. They also would have to deal with a campaign of misinformation and an attempted cover-up of important findings on the part of the radium company.

Important evidence and critical findings would be hotly debated and disputed. In time, the debate would become so intense it would erupt into an extremely bitter controversy in the nation's medical literature. Individual investigators, and their findings, would be severely attacked and ridiculed by physicians, scientists, pharmaceutical companies, and the radium industry.

To finally solve the mystery, additional cases of the new disease would have to be found at other dial-painting facilities in Waterbury, Connecticut, and Ottawa, Illinois, thus proving the sickness was not caused by some unique factor which only occurred at the radium plant in Orange, New Jersey. Eventually, the new occupational disease of radium poisoning, a form of chronic radiation sickness, would be recognized.

The first victims of radium poisoning would die from aplastic anemia and related complications, while later victims would succumb to rare radium-induced head and bone cancers and sarcomas (Figure 2). Although many of the deaths would occur in the 1920s and 1930s, others would die decades later. The last death occurred in 1988. In total, 112 radium dial workers are known to have died from the occupational disease.[10]

In time, the tragedy of the radium dial workers would be recognized as the world's first mass experience with injury and death caused by exposure to atomic radiation. Eventually, the dial workers would gain an important place in history as the world's first known victims of the dawning atomic age. The tragedy would be an important turning point in modern history. It would mark the beginning of the end of the world's enchantment with and unbridled enthusiasm for radioactivity. The tragedy also would serve as an important warning of the perils of the careless, inappropriate, and unwise use of radioactivity. The warning, however, would only be slowly and reluctantly heeded by the medical profession, the scientific community, and the general public.

Starting in the mid-1920s, the radium dial workers became famous throughout the world. The tabloid papers frequently referred to them as the "living dead," the "walking ghosts," and the "doomed women."[11] Year after year, the newspapers reported their slow, agonizing deaths. As a result of the publicity, some of the women received worldwide sympathy and support. Yet, despite the enormous public concern, they were of interest to only a handful of researchers.

During World War II, however, scientific interest in the former workers greatly increased. The occupational exposure standard developed for the radium dial painters would become the primary safety standard for the U.S. atomic-bomb-producing Manhattan Project. To the top secret bomb project, the terrible radium dial tragedy would prove to be truly "a most valuable accident."[12] If it had not occurred, tens of thousands of defense workers might have been in great danger.

During the Cold War, the nation and the world would again turn to the radium dial workers. This time the women would provide unique and invaluable information on the possible long-term health effects of radioactive fallout from aboveground nuclear testing. For decades, the U.S. Atomic Energy Commission, and later its predecessor, the U.S. Department of Energy, would conduct an intensive search to find and study as many of the workers as possible. Thousands of dial painters would be located, interviewed, examined, and monitored to identify changes in their health status. In addition, the remains of many dial painters would be exhumed, and the amount of radium in their bones measured. Ultimately, the studies of the radium dial workers, along with those of the Japanese atomic bomb survivors, would form the basis of much of the world's present knowledge of the health risks of radioactivity.

Today, the story of the radium dial tragedy is considered a classic tale in the fields of occupational medicine, health physics (radiation safety), and public health. Even though it is a classic, however, many do not know the entire story. And most of the general public is totally unaware of this important saga.

The tragic and heroic story begins with the discovery of radium.

Chapter 1

Dawn of a Miracle

The scientific history of radium is beautiful.[1]
Marie Curie, 1921

Brilliant Discoveries

In 1898, Pierre and Marie Curie were intensely studying the uranium ore pitchblende. Two years earlier, the French physicist Henri Becquerel (1852–1908) accidentally found that uranium spontaneously emitted mysterious rays, similar to x-rays. Both the Polish-born Marie Sklodowska Curie (1867–1934), who was searching for a topic for her doctorate degree, and her French physicist husband, Pierre (1859–1906), were excited by the new phenomenon (Figure 3). Because the discovery was so recent and so little known, the couple decided to investigate it. To describe and define this new property of matter, Marie proposed to call it radioactivity.[2]

During the course of her investigation, Marie Curie found that pitchblende ore was several times more radioactive than expected, based on its uranium content alone. From this observation, she concluded that the ore must contain some unknown chemical element more radioactive than uranium. Marie further reasoned that the element must exist in only infinitesimal quantities, or else it would have been previously detected, and that it must be powerfully radioactive.

With the help of her husband, Marie began to systematically search for the unknown element. After six months of chemically separating the various components of pitchblende and measuring their radioactivity, the couple finally found what they were looking for: a new element. In honor of Marie's native country, Poland, they named it polonium.[3]

The Curies continued their research, and after several more months, they discovered a second new radioactive element. However, this element occurred in such small quantities it could only be detected as a trace impurity. Late in December of 1898, the couple announced their discovery.[4] This element, which would become much more famous than the first, they named radium, or the giver of rays.[5]

Figure 3. *Marie and Pierre Curie.*

To prove conclusively the existence of radium, the Curies had to isolate a pure sample of the extremely rare element. This would prove to be extraordinarily difficult. As Marie later wrote, it would be "like creating something out of nothing."[6] The couple was extremely handicapped in their endeavor. They had little money, they did not have an adequate laboratory, and they had no supplies or personnel to help them. The only thing they did have was enormous perseverance and determination.

To help the couple, the Paris School of Industrial Physics and Chemistry, where Pierre worked, gave them some basic laboratory equipment, and permitted them to use an abandoned shed. This dilapidated, old wooden structure, which the Curies turned into a makeshift laboratory, had a cinder floor, high plastered walls, and a skylight that did not keep out the rain. It was furnished with only a few worn pine tables, a small inadequate cast-iron stove, and a blackboard.

The Curies would find the old shed suffocatingly hot and dusty in the summer, and bitterly cold and damp in the winter. Its humidity would pose serious problems for their delicate instruments. And its dust would interfere with their intricate crystallizations. Moreover, the shed lacked exhaust hoods to remove the toxic fumes produced by the numerous chemical separations the couple would undertake.

One chemist who eagerly visited the Curies' laboratory to see the birthplace of radium was shocked to find it "a cross between a horse stable and a potato cellar." He further commented that: "if I had not seen the worktable with the chemical apparatus, I would have thought it a practical joke."[7]

To obtain sufficient quantities of pitchblende, the Curies appealed to a colleague who was a member of the Academy of Sciences of Vienna. With his help, the Vienna academy requested the Austrian government, which owned uranium mines in St. Joachimsthal, Bohemia (now Jachymov in the Czech Republic), to supply them with the necessary ore. The director of the mines agreed to give the couple the worthless pitchblende residue for free, providing they pay the cost of its freight.

After receiving their first ton of ore, the Curies started the laborious series of chemical separations necessary to concentrate and isolate the radium. The ore would have to be crushed, pulverized, leached, precipitated, and crystallized. The couple divided their labor. Marie would conduct the separations, while Pierre would measure the radioactivity of the various components.

Marie spent days mixing and stirring a boiling cauldron with a large, thick iron bar which was about as tall as she was. Other days, she worked on minute and delicate crystallizations. Marie would repeat this exhausting process over and over again, month after month, year after year. Finally, after nearly four years of unremitting toil, the couple produced less then a thimbleful, one-tenth of a gram, of nearly pure radium salt.

Pierre Curie had hoped radium would have a beautiful color. Unfortunately, the element was a dull, metallic white. Nevertheless, radium had one unhoped for characteristic. To the Curies' great amazement and delight they found that it *glowed*.

Marie described the couple's feelings:

> Sometimes we returned in the evening after dinner for another survey of our domain. Our precious products, for which we had no shelter, were arranged on tables and boards; from all sides we could see their slightly luminous silhouettes, and these gleamings, which seemed suspended in the darkness, stirred us with ever new emotion and enchantment.[8]

Soon after its discovery, radium was found to have powerful physiological effects. In 1900, a German researcher reported that he developed skin burns a few days after handling tubes of radium. The researcher also indicated that he tested the effects of the new element by purposely exposing his arm to it.

After reading the report, Pierre Curie decided to test radium's effects himself. He took a small amount of the element and strapped it to his arm and left it there for 10 hours. When he removed it, Pierre found the area under the radium had turned red. Over the next several days, this turned into a wound that took 4 months to heal. Pierre repeated the test, exposing his arm for half an hour, and for eight minutes. Each time a wound developed and then healed.[9]

Radium's physiological effects also were accidentally discovered by Henri Becquerel. After borrowing a tube of radium from the Curies, Becquerel carried it in his vest pocket for several hours. Later he was surprised to find a sore appear at the exact spot where the radium had been. Becquerel was both annoyed and delighted by his discovery. He told the Curies of "this evil effect of radium," saying of the new element: "I love it, but I owe it a grudge!"[10]

These discoveries led to radium's application in medicine, where it was used to destroy diseased cells. To develop new medical uses, Pierre Curie frequently worked with physicians, loaning them samples of radium.

After radium's discovery, it quickly became a sensation. Everyone was fascinated by the mystery and romance of the new element. To science, radium was revolutionary. Here was an element that spontaneously released energy; that continuously glowed; and that slowly transformed itself into other elements, a sort of alchemist's dream come true. One 1903 magazine article summed up radium's importance by declaring that it was to modern scientific research what the invention of the telescope was to astronomy.[11]

The discovery of radium would profoundly change basic concepts about matter and the atom itself. The study of radium would lead to an important new branch of science, nuclear physics.

Radium was equally revolutionary in the field of medicine. Initially, the element was used to treat lupus and a few other skin diseases. In time, however, it would be used to treat a multitude of other diseases. Eventually, radium therapy, or Curie therapy as it was called in France, would become an important new branch of medicine, nuclear medicine.

To the general public, radium seemed magical. The popular press as well as numerous scientists, physicians, lecturers, poets, and novelists made wild and fantastic claims as to what radium could and would do in the future. It was claimed, for example, that the energy from radium would replace natural gas and electricity. Radium paint, instead of electricity, would be used to illuminate rooms. Some claimed that radium would render future warfare impossible because its heat rays would destroy fortifications and crumble warships. It would supposedly make the blind see, the deaf hear, cure insanity, and even determine the sex of unborn children. Others claimed the miracle element was the legendary fountain of youth. A few even claimed that radium was the source of life itself (Figure 4).[12]

There was so much speculation that newspapers and journals began cautioning their readers to use common sense. One medical journal, for example, wrote:

> The real properties and the alleged properties of radium would seem to have overturned the rational reasoning faculties of a large portion of the American public.... Statements for which there is no foundation whatever are being constantly advanced as to the marvelous events that can and will be brought about through the influence of radium.[13]

Figure 4. *Newspaper illustration entitled "Hope" from the* Page Los Angeles Examiner, *1914.*

The year 1903 was an important one for the Curies. Marie finally completed her thesis and was granted a doctorate degree, the first ever given to a woman by the University of Paris, the Sorbonne. Her thesis, *Research on Radioactive Substances*, quickly become a classic in science. Within a year, it was printed in many editions and languages, and it was serialized in numerous scientific journals.[14]

At the end of the same year, the Curies, along with Henri Becquerel, were jointly awarded the Nobel Prize in physics. Becquerel was given the prize for his discovery of radioactivity, while the Curies received it "in recognition of the special service rendered by them in the work they jointly carried out in investigating the phenomena of radiation...."[15] Marie Curie was thus the first woman to win the prestigious prize.

Ultimately, the award would change the Curie's lives forever. The modest and private couple became instant celebrities, and the unwanted fame was overwhelming. They were inundated with visitors and requests for articles and lectures, and they were hounded by journalists, photographers, autograph seekers, socialites, and an adoring public.

In 1906, Marie Curie's life was again changed dramatically, this time by a dreadful catastrophe. Her husband, Pierre, was instantly killed in a traffic accident. His death left her devastated. Henceforth, she was alone not only to raise the couple's two young children, but to continue their research.

After Pierre's death, the University of Paris decided to break its long tradition of a male-only faculty and asked Marie to succeed her husband. She would become the first woman professor in the university's 650-year history.[16]

In 1911, Marie again received an exceptional honor. She was the first Nobel laureate to win the prize twice. This time, she was the sole recipient in the field of chemistry. Her second Nobel was given:

> ... in recognition of the services rendered by her to the development of chemistry: by her discovery of the elements of radium and polonium, by her determination of the nature of radium and isolation of it in a metallic state, and by her investigations into the compounds of this remarkable element.[17]

Marie Sklodowska Curie would be proclaimed the world's greatest woman scientist. And the story of her discovery of radium would become one of the most publicized and popular sagas in the history of science.

Few Warnings, Few Precautions

Initially, the Curies were completely unaware of the dangers of radium. At first, the quantities of radioactive material they worked with were small, weak, and posed little risk. However, as the couple continued their work, they produced larger and much more powerfully radioactive samples of radium.

As they worked, the Curies were constantly exposed to radium's harmful rays. They regularly ingested and inhaled radioactive particles and dust, and they frequently breathed high concentrations of radon, the radioactive gas given off by radium. From this exposure, the Curies slowly began to experience the damaging effects of radioactivity.

Marie's fingertips hardened, became red, and were very painful from handling containers of radium. The inflammation would last for weeks, and the skin of her fingertips would peel off. As new skin grew back, however, her fingertips would remain painfully sensitive. She also suffered terrible burns on her hands. These took a long time to heal and left permanent scars. In addition, Marie constantly felt weak and fatigued. And she suffered from chronic anemia.

Pierre's hands, like his wife's, also would be injured from handling radium. For a time, they would be so severely impaired that he would have trouble dressing himself. Pierre would also experience debilitating pain in his legs and back; sometimes it was so intense that he could not get out of bed. His malaise was so severe and lingering that he had to postpone for nearly two years his trip to Sweden to deliver the couple's Nobel lecture.

The Curies eventually recognized that radium, and radon, could be harmful or even deadly. Pierre conducted several experiments that clearly demonstrated the danger. In one experiment, he found that a few milligrams of radium implanted near the vertebral column of a mouse produced death by paralysis within three hours. In another, he placed tubes of radium in contact with the back of the necks of guinea pigs. The animals were paralyzed or died within a few hours, depending upon their length of exposure.[18]

Pierre would tell a visiting American engineer, William J. Hammer, of the dangers of radium indicating: "that he would not care to trust himself in a room with a kilo of pure radium, as it would burn all the skin off his body, destroy his eyesight and probably kill him."[19]

Pierre Curie also would underscore radium's harmful effects in the couple's Nobel lecture, stating:

> It is possible to conceive that in criminal hands radium might prove very dangerous, and the question therefore arises whether it be to the advantage of humanity to know the secrets of nature, whether we be sufficiently mature to profit by them, or whether that knowledge may not prove harmful. Take, for instance, the discoveries of Nobel—powerful explosives have made it possible for men to achieve admirable things, but they are also a terrible means of destruction in the hands of those great criminals who draw nations into war. I am among those who believe with Nobel that humanity will obtain more good than evil from future discoveries.[20]

Ironically, however, despite their knowledge of its harmful effects, the Curies would refuse to admit that the illnesses from which they chronically suffered were caused by radium. Instead, the couple would dismiss or shrug them off, or they would blame them on such things as overwork, poor diet, or rheumatism from working in their damp laboratory.

The Curies also refused to take safety precautions to protect themselves, and others, from radium's effects. Marie Curie would never give any special warnings to the many researchers who flocked to her laboratory to work and study. And they would be taught few precautions. One researcher, for example, was told only to change his laboratory coat frequently![21] When two of Marie's assistants suddenly became sick and died from their exposure to radium, she tended to blame them for not getting out in the fresh air more frequently.

Throughout her life, Marie Curie never admitted that radium, the precious element that she and her husband discovered and so painstakingly isolated and concentrated, which was curing cancer and saving lives, was slowly killing her. However, on July 4, 1934, at the age of 66, she died from aplastic anemia, her bone marrow destroyed by her many years of exposure to radium.[22]

Unfortunately, because there were so few warnings, and only a small number of persons were exposed to the new element for long periods of time, for decades radium would enjoy the false reputation of being harmless.

Chapter 2
The First Nuclear Industry

We have an ore supply in carnotite, so we will make it ourselves.[1]
Joseph M. Flannery, 1910

Radium Comes to America

Soon after its discovery, radium was brought to America. In 1899, physics professor George F. Barker (1835–1910) of the University of Pennsylvania became the first person to exhibit radium in the United States. Barker, who purchased a small sample of the new element from a German firm, was very enthusiastic about its possible medical uses. He felt that radium would prove to be much more convenient and economical than x-rays for surgical exploration.[2]

Radium also was brought to America by William J. Hammer (1858–1934) (Figure 5), an electrical engineer and former assistant to the great inventor Thomas A. Edison. In 1902, while in Europe, Hammer visited the Curies. With their help, he purchased nine tubes of the new element from a French company.

When he returned to the United States, Hammer immediately began experimenting with radium. He mixed some of the element with gum damar and produced the world's first radium paint. He applied the glowing paint to watch and clock dials, gunsights, escutcheon plates of keyholes, poison bottle labels, and a number of other items. Hammer, however, did not attempt to patent the paint because the new element was "worth $36,000,000 a pound ... and there were only a few grams of radium in the whole world."[3]

Hammer also gave his former employer, Thomas Edison (1847–1931), some radium to experiment with (Figure 6). The great inventor used the element to try to discover new rays. In addition, Edison attempted to identify chemicals that glowed when exposed to the element, finding more than 100 of them, some of which glowed very strongly. And he attempted to test radium's biological effects by exposing insects and plants to the element.[4]

Edison, who had previously been seriously injured while experimenting with x-rays, realized that radium's rays could also be very dangerous. He repeatedly warned the

Figure 5. William J. Hammer.

public and others to be very cautious with the new element. In a newspaper interview, he admonished those using it, stating:

> Is it not reasonable to say that the use of radium, which is much more powerful than the x-ray, will effect the body quicker and do more harm? Remember, I am not trying to discourage experiments. That is farthest from my mind. But I do caution those who use the substance to exercise care, both for their own sake and for the sake of their patients. It took centuries to develop electricity upon its present scale, and it may take years for us to get any definite ideas about radium.[5]

Figure 6. *Thomas Edison at his laboratory in West Orange, New Jersey.*

Despite his concerns, Edison, like Becquerel, unwittingly carried some radium in his vest pocket for several days. Aware of its possible harmful effects, Edison's family and staff feared that he would be seriously injured. He assured them and the public, however, that nothing was wrong.

In an interview concerning the episode, Edison again took the opportunity to warn others of radium's danger. He declared:

> I believe that we should first learn all about radium, and then go to the matter of applying it for the good of humanity, but great care should be taken at all times. There may be a condition into which radium has not yet entered that would produce dire results, and everybody handling it should have a care.[6]

For a time, Edison was very interested in trying to find a domestic source of radium. Using the vast collection of material at his enormous West Orange, New Jersey, laboratory he tested numerous rock and mineral specimens to see if they contained the new element. However, after many unsuccessful attempts he eventually abandoned his search.

America's Radium Fields

As radium's fame grew, many scientists and physicians, as well as numerous medical centers, universities, and commercial companies, all wanted to purchase the element. Demand for radium, however, far exceeded supply, and its price skyrocketed. The rare element quickly became the most precious and valuable substance on earth.

There was such a great demand for radium that it appeared that the supply of pitchblende from the St. Joachimsthal mines in Bohemia, which constituted almost the entire world's supply of the element, would soon be exhausted. To conserve the rare and valuable substance, at the end of 1903, the imperial Austrian government placed an embargo on the export of all pitchblende ore and residue. Under its strict control, only a very small, select number of persons and institutions were able to obtain the radium-bearing ore.

Many scientists and physicians deplored the embargo; without a ready supply of the new element they would be unable to obtain its wonderful benefits. The dean of the medical faculty of Flower Hospital in New York City, for example, declared:

> Further progress in the use of radium for curing diseases will be practically impossible.... When Prof. Curie and other eminent European scientists are totally unable to procure desirable specimens of this substance, there is small chance of any one else doing so. The Austrian Government has positively refused to allow any more of it to leave that country for the present, and there is as yet no other known source of what may be called a working supply of the element.[7]

To meet the enormous demand, prospectors searched the world over for additional sources of supply. Soon new deposits of radium-bearing ores were discovered in Portugal, Madagascar, Hungary, Norway, Sweden, Russia, Canada, and Australia. The largest of all, however, were found in America, in southwestern Colorado and southeastern Utah. These deposits were carnotite ore.

Particularly large reserves of this ore occurred in Paradox Valley in Montrose County, Colorado. The valley, which is approximately 20 miles long and 3 miles wide, was so named because the Dolores River cuts directly across it, instead of running parallel to it (Figure 7). This isolated and remote area, which only a few years earlier had been known for outlaws, cattle rustling, and gunfights, would become America's most important radium field.[8]

Carnotite, a bright canary-yellow mineral, is a chemically complex ore containing potassium, vanadium, and uranium. Soft, powdery, and heavy, it is found widely scattered in pockets of sandstone, limestone, and clay. Although it contains radium and deposits are vast, it is a very low-grade ore, containing only approximately one part radium for every 180,000,000 parts of ore.[9]

Carnotite is believed to have been first used as a yellow pigment by the Ute and Navajo Indians long before the first white men came to the region. Early settlers noted the existence of the brightly colored sandy mineral. In 1896, a sample of the ore was

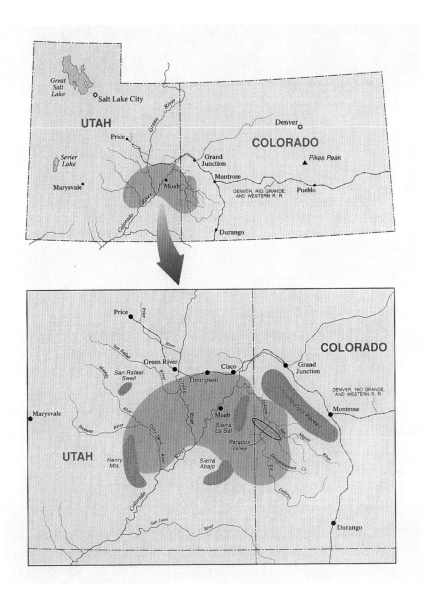

Figure 7. Area of major carnotite ore deposits in Colorado and Utah.

sent to the Smithsonian Institution in Washington, D.C., for analysis, where the specimen was found to contain uranium.

Not until 1899, however, was carnotite first identified as a new mineral. Several years earlier while visiting Colorado in search of rare metals, French mineralogists Charles Poulot and Charles Voilleque found the ore. Not knowing its content, they sent samples of carnotite to the French Academy of Sciences in Paris, where the chemical composition was determined. Because the mineral had not been previously identified, carnotite was named in honor of the famous French mining engineer and chemist, Marie Adolphe Carnot (1839–1920).[10]

Figure 8. Stephen T. Lockwood.

The discovery of the radium-bearing ore created a sensation in Europe. Many prospectors, chemists, and promoters were sent to America to purchase as much of the mineral as possible. And several British, French, and German firms established companies to mine the ore for shipment to Europe to have the uranium and radium extracted. By the early 1910s, so much of the ore was being shipped abroad that the Chief of the Division of Mineral Technology of the U.S. Bureau of Mines charged: "The uranium deposits of Colorado and Utah are being rapidly depleted for foreign exploitation, and it would seem to be almost a patriotic duty to develop an industry that will retain the radium in America."[11]

The first person to attempt the commercial production of radium in the United States was Stephen T. Lockwood (1875–1971) of Buffalo, New York. Lockwood, a young Princeton-trained lawyer, had a strong interest in mine engineering (Figure 8). While traveling in Utah in 1900, he met two prospectors, John Walsh and James H. Lofftus, who had discovered a large deposit of carnotite. They asked the young East Coast lawyer to find out what the ore could be used for, and whether the mineral was marketable. Lockwood believed that the uranium in the ore might have commercial value, so he encouraged the prospectors to stake out claims.

Upon returning to Buffalo, Lockwood subsequently became interested in radium. He thought that carnotite might contain the newly discovered element, and in 1902, he began to experiment with the ore to see if he could extract its radium. He chemically removed the ore's uranium and vanadium and was left with a residue of fine white sand. To see if it contained radium, he attempted to create a radiograph. Lockwood placed the residue along with a key and several coins in a rubber container and then placed it on a photographic plate. When the plate was developed, the image of the key and the coins were clearly visible.[12]

Lockwood also sent a sample of the ore to Alexander H. Phillips (1866–1937), professor of mineralogy at Princeton University, to see whether radium could be isolated from it. Late in 1902, after a month of experimentation, Phillips produced a specimen of the new element. He described the chemical process he used in an article entitled "Radium in an American Ore." In the article's concluding sentence, Phillips confidently predicted: "These facts prove beyond question that carnotite will become a commercial source of radium."[13]

As soon as Lockwood was sure that carnotite contained radium, he sent a sample of the ore to the Curies, and he boldly wrote them, asking that they identify the best commercial method of recovering the element from the ore. The Curies, who believed that it was not in the proper scientific spirit to materially profit from their discovery of radium and who never attempted to patent the process they used to extract the element, wrote back to Lockwood providing him with detailed information.

In May of 1903, Lockwood incorporated the Walsh-Lofftus Uranium and Rare Metals Company (later renamed the Rare Metals Reduction Company) in Buffalo. During the same year, he established a reduction plant at nearby Lackawanna, New York.

Lockwood planned that his company would mine about 300 tons of carnotite ore each year from its claims in Utah. The ore would then be shipped across country by rail to the Lackawanna plant for processing. At the reduction plant, which was designed to treat at maximum capacity two tons of carnotite per day, the ore's uranium, vanadium, and radium would be removed. Each element would then be sold separately at a profit.

In 1908, after several years of operation, the firm closed. It could not obtain an adequate supply of ore, there was little economic demand for uranium and vanadium, and its efficiency of radium recovery was just too poor. As a result, the company never produced any commercial radium.[14] Although Lockwood's company ended in failure, it did, however, succeed in making other American corporations aware of the vast deposits of radium-bearing carnotite ore in the West.

Corporate Giants

By all accounts, Joseph M. Flannery (1867–1920) and his brother James J. Flannery (1854–1920) were remarkable businessmen. Both men had strong determination and broad vision. Joseph, who was more than ten years younger than his brother, was at-

Figure 9. *Joseph M. Flannery (fourth from left), James J. Flannery (fifth), and Henry Ford (seventh).*

tracted by high risk and adventure (Figure 9). He had outstanding ability to convince and convert others to the practicability of his dreams. He also was a promotional and marketing genius. James, on the other hand, was quiet and conservative. His strengths were in planning, organizing, and administering complex operations and businesses. He also was a shrewd banker and a financial genius.

The brothers, who had little formal education and no experience whatsoever in mining or industry, started their careers as undertakers, running a very successful funeral practice in Pittsburgh for a number of years. However, in 1904, they decided to leave the business. After selling their interests, they used their capital to purchase the patent of a special bolt designed to hold together the fireboxes of steam locomotives. Unlike other bolts, which tended to break from the extreme heat and pressure, the Tate Flexible Staybolt expanded and held firm. To manufacture and market the new bolt, the brothers founded the Flannery Bolt Company of Bridgeville, Pennsylvania.[15]

In seeking the best steel to make their bolts, Joseph taught himself metallurgy. To learn as much as he could, he traveled to the leading steel centers of the world to study the latest developments and techniques. While in Europe, he found that by adding traces of the element vanadium to steel, it could be made stronger, more elastic, and more durable. However, because there were no large, rich deposits of vanadium, the element was not widely used commercially.

In 1905, the Flannery brothers learned that some mines in Peru contained deposits of vanadium ore. The brothers sent an electrical engineer and a mining engineer to investigate. While in Peru, the engineers went to several sites that had vanadium, but the ores were very poor quality. Just before they were to leave the country, however, they heard a rumor that a large deposit of vanadium had been found. They contacted the owner of the deposit, Eulogio Fernandini, who gave them permission to visit it.

After much difficulty, the engineers reached the remote and desolate site, which was located high in the Andes Mountains. At the site, they were astonished to find enormously rich deposits of the ore. Vanadium seemed to be everywhere; even the soil was stained with the various hues of the element.

When they returned, the engineers immediately asked Fernandini to grant them an option on the property. However, he refused, saying he first wanted to have the deposit fully evaluated. The engineers did, however, convince him to agree not to do anything until they returned from the United States.

When Joseph Flannery heard the news, he immediately left for Peru. On arriving there, he found that a British company was also very interested in developing the deposit. If he was to obtain the mineral rights to the property, he would have to act quickly and decisively.

Flannery took all the money he had with him and converted it into gold pieces. He placed the coins in an old carpetbag. And when he met Fernandini, he poured the glittering mass of coins on a table in front of him. Telling the owner of the property: "Here's $20,000 in gold," Flannery then divided the coins in half, saying: "This pile goes to you to pay for all your rights; the other pile is to be used for the development of the mine." This dramatic gesture won over Fernandini, who quickly accepted the offer.[16]

When the site was evaluated, it was found to contain the world's largest and richest deposits of the ore, and for decades, it would supply the vast majority of the world's vanadium.

Although the Flannery brothers now had a large source of vanadium, they were faced with the difficult task of convincing American companies to adopt it. Joseph Flannery would single-handedly undertake the task of promoting and marketing vanadium to the nation's industry. His campaign would prove to be both dramatic and highly successful.

One of Flannery's successes was persuading Henry Ford (1863–1947) to use the alloy for a new line of cars he was introducing—the Model T. To convince the automaker, he invited him to Pittsburgh to witness a dramatic test of the alloy's strength. Flannery had a special automobile frame built of vanadium steel. Then he took Ford up to a high cliff, where he had the frame thrown over the precipice. It fell hundreds of feet and landed with a resounding crash. When Ford examined the frame, he was astonished to find that although it was twisted and bent, it did not break.[17]

The demonstration worked. Ford decided to use the alloy to make numerous parts of his new cars, including their crankshafts, drive shafts, gears, rods, and springs. In all, half of the steel used in the cars would be made with the new alloy.[18] Vanadium steel would make them lighter, stronger, and cheaper to mass-produce, and the alloy would greatly contribute to the Model T's enormous success by making the cars more durable and reliable.[19] Soon, all of the other automakers in the nation began demanding vanadium steel.

In a short time, the Flannery brothers became wealthy. Their staybolt was used by almost all of the locomotives in America, and by many railroads in other countries. They controlled the world's largest and richest deposit of vanadium, and demand for the element was enormous. Their future seemed assured.

A personal tragedy, however, would change the brothers' lives forever. In 1908, their sister, Eleanor Flannery Murphy (1856–1910), was diagnosed with terminal cancer of the uterus. Joseph, who had always been very close to her, was devastated. He asked her physicians what could be done to save her. They said, with regret, they could do very little. However, they did indicate that radium might perhaps help; in Europe, radium was reported to have cured many types of cancer. Unfortunately, the doctors also said that only a handful of physicians and hospitals in America had a supply of radium, and no one in Pittsburgh had any of the rare element.

In a heroic attempt to save his sister's life, Joseph went to Europe to obtain radium himself. Money was no object. He would pay whatever was necessary. After months of searching, however, he found no one willing to sell him a sufficient quantity of the element. Failing in achieving his goal, he returned home to be with his sister as she slowly died from her insidious illness.[20]

After her death, Joseph Flannery vowed he would find a cure for cancer. Together with his brother James, he would organize and incorporate the Standard Chemical Company of Pittsburgh for the commercial production of radium. Their plan was to extract the element from the carnotite ore of Colorado and Utah. The brothers hoped that their new company would become to radium what John D. Rockefeller's Standard Oil Company was to oil—the world's largest producer.

Joseph Flannery would turn all of his attention to developing and running the new company. It would be a tremendous gamble, but he would risk everything in his effort to produce the radioactive element.

Many of his friends thought he was foolish, and his bankers believed he was seriously jeopardizing his wealth. He had no scientific or technical knowledge of radium. The Western carnotite ore contained very little radium, and there was no known commercial process to extract it. Further, even if he was lucky enough to produce the element, there was no market for it in America.

Despite the obstacles, Flannery would use his personal fortune to purchase mining claims, milling and mining equipment, and an old stove factory in nearby Canonsburg, Pennsylvania, which he converted into a large radium extraction plant. He also would hire hundreds of prospectors, miners, processors, and a team of chemists.

Figure 10. A Colorado radium mine.

He would continue to face many difficulties in the years ahead. Mining the ore and chemically processing it would prove to be enormously costly, and difficult. First, the ore had to be mined and sorted by hand (Figure 10). And for every six to ten tons of ore mined, only about one ton was suitable for shipment. Next, the ore was put into bags weighing 70 to 80 pounds each and transported by burro as much as 25 miles down steep, narrow, winding paths to a local mill. At the mill, the ore was crushed and concentrated before it was sent by horse-drawn wagon 65 miles to Placerville, Colorado. There it was loaded onto a narrow gauge railroad and sent to Salida, Colorado, where it was transferred to the transcontinental railroad. From Salida, it traveled 2,300 miles to Canonsburg, Pennsylvania, to the company's refining plant.[21]

At the refining plant (Figure 11), the ore went through a laborious chemical process, which took several chemists (Figure 12), working by trial and error, more than 18 months to perfect. To produce just one gram of radium—a mere one-twenty-eighth of an ounce—would ultimately require 500 tons of carnotite ore, 500 tons of chemicals, 10,000 tons of distilled water, the energy of 1,000 tons of coal, and the labor of 150 men for one month. In addition, it would require the effort of 15 chemists conducting numerous crystallizations, control tests, and analyses for five weeks to complete the process.[22]

Figure 11. Standard Chemical Company's radium refinery at Canonsburg, Pennsylvania.

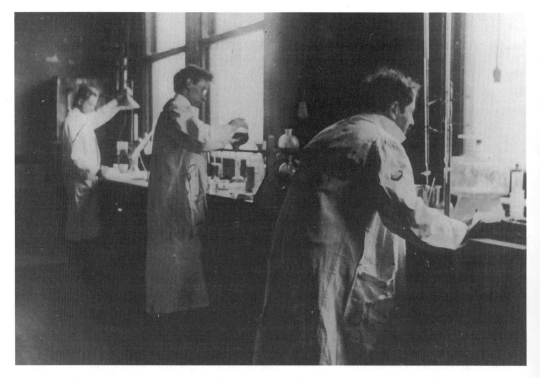

Figure 12. Chemists at the Standard Chemical Company, Pittsburgh.

In 1913, after three years of effort, the company produced its first radium. However, at the time there were few hospitals or American physicians who understood and appreciated the value of the costly new element. Lacking a domestic market, the company sold its entire output of radium, worth approximately $1 million, to various European countries, principally Germany.[23]

Shortly after selling the radium, Flannery was faced with two other very serious threats: competition from a formidable new company, and possible nationalization by the federal government.

In 1913, the U.S. Bureau of Mines, which was very concerned about the amount of carnotite ore being shipped out of the country, requested that Dr. Howard A. Kelly (1858–1943) and James Douglas (1837–1918) create a radium institute for the treatment of diseases, similar to those in Europe.

Dr. Kelly, a cancer specialist and faculty member of the Johns Hopkins Medical School, was one of the first physicians in America to use radium, ironically purchasing it from the Standard Chemical Company. Specifically, he wanted to obtain a large supply of the element at a low price, for his own private clinic in Baltimore.[24] Douglas, on the other hand, was a mining engineer, industrialist, and philanthropist, who wanted radium for the General Memorial Hospital of New York City, where he was a trustee.

Kelly and Douglas agreed to work together, and in September of 1913, they incorporated the not-for-profit National Radium Institute. Several weeks later, the bureau signed a cooperative agreement with the institute to form a joint business-government partnership to produce radium.

The institute agreed to spend $150,000 by August 30, 1916, to purchase carnotite ore and to construct a radium processing plant, while the bureau would provide the necessary technical support and supervision of the mining and processing operations. In return, Kelly and Douglas would obtain the radium produced by the institute, and the bureau would gain invaluable experience and knowledge of the methods of processing and producing radium, which it could freely disseminate to others.

The institute leased carnotite claims; a processing plant was constructed in Denver, Colorado; and the bureau's technicians and scientists began experiments to develop the most cost-effective method of extracting and processing radium.[25]

At the beginning of 1914, Joseph Flannery would have to face an even more devastating threat. That year, the U.S. Secretary of the Interior Franklin K. Lane (1864–1921), a conservationist who wanted to maintain the natural resources of the West, had several bills introduced in Congress which, if passed, would force the Standard Chemical Company to go out of business.[26] The sweeping legislation would nationalize all carnotite, pitchblende, and other radium-bearing ores on public lands, provide for the purchase of these ores by the federal government, and authorize the construction of a radium processing plant which would be run by the U.S. Bureau of Mines.

The House and Senate Committee on Mines and Mining held hearings on the bills in Washington, D.C.[27] And for a short time, it appeared that they might pass. However, Flannery, other mining companies, and state and local politicians in Colorado mustered massive opposition. In the end, the government's effort to control the nation's radium fields and to create a monopoly died in Congress.

Although the government failed to take control of radium, the joint effort of the U.S. Bureau of Mines and the National Radium Institute was highly successful. By the end of 1916, the institute had produced several grams of radium at an average cost of only $39,325 per gram, one-third of the world market price of $120,000. Having produced a sufficient quantity of radium, the Denver plant was closed in 1917, and the institute was dissolved.[28]

From 1913 to 1922, the United States was preeminent in the world's radium market, producing four-fifths of the world's radium. However, its domination quickly ended when Belgium began producing radium from very high-grade pitchblende ore mined in the Katanga region of the Belgian Congo (now the Democratic Republic of the Congo). America's low-grade carnotite ore just could not compete, and within a few years, almost all of the nation's radium mines and processing plants closed.[29]

During America's radium boom, the Standard Chemical Company was by far the world's largest radium producer. In all, it produced 74 grams of radium, accounting for almost half of the nation's total production.[30] In 1920, the company reached its peak production, producing 18.5 grams of radium. That same year, within several weeks of each other, Joseph and James Flannery died: victims of the Spanish influenza epidemic and overwork.

Unfortunately, Joseph Flannery did not live to receive what would have been his greatest honor. In 1921, when Marie Curie came to America for the first time, she spent several days touring the Standard Chemical Company's Pittsburgh headquarters, research laboratories, and Canonsburg refining plant. During her visit, she was given a gram of radium, which the company produced and specially packaged for her. President Warren G. Harding (1865–1923), in a White House ceremony attended by many dignitaries, gave the great scientist ownership of the radium, which had been purchased for her on behalf of the women of America (Figure 13). Joseph Flannery was sorely missed at the ceremony.[31]

At the end of 1922, when it was clear that the Belgians were going to gain monopoly control of the radium market, the Standard Chemical Company entered a contract with Radium Belge to become its sole representative for the Western Hemisphere. In return, the company agreed to cease all mining and refining operations and to exclusively market the Belgian radium for five years. When the contract expired in 1927, however, the Belgians refused to renew it. And with the help of several former directors of the Standard Chemical Company, they entered the American market, forcing the company to close.

Figure 13. Marie Curie and President Harding outside the White House.

With the Belgian monopoly, all of America's other radium producers quickly stopped mining and processing the element, except the U.S. Radium Corporation of Orange, New Jersey. That company continued to mine and process carnotite ore for several more years. However, even it eventually stopped production and purchased its supply of radium from the Belgians.[32]

Chapter 3

Radium Medicine

At present, we can safely say that radium salts introduced intravenously ... are perfectly harmless. Thirty patients given one, and sometimes several injections, did not show the slightest alarming symptom or any injurious effect to any organ. This observation extended over a period of from three to six months.[1]

Dr. Frederick Proescher, 1914

External Use

One of the first to investigate the possible medical uses of radium was Dr. William Rollins (1852–1929), a Boston physician and dentist. Using a sample of radium he obtained from George Barker, Rollins in 1900 attempted to find a substitute for x-rays. Several years later, Rollins also tried to interest other physicians in experimenting with radium.[2]

In 1901, French physicians at the St. Louis Hospital in Paris made use of radium loaned to them by Pierre Curie in an effort to cure lupus and other skin lesions.[3] The physicians, who treated patients by holding an ampule of radium over the afflicted area, found that in many cases the element was remarkably effective (Figure 14). The penetrating radiation given off by the radium destroyed the diseased area, enabling healthy cells to grow back.

Later, radium was employed in the treatment of various types of cancers. At first, it was used on skin and other external cancers. In time, radium also would be used to treat cancers deep within the body.

One of the first to suggest a method of applying radium to these cancers was Alexander Graham Bell (1847–1922), the inventor of the telephone. In 1903, Bell wrote to a New York City physician, suggesting:

> I understand ... that the Roentgen rays [x-rays], and the rays emitted by radium, have been found to have a marked curative effect upon exter-

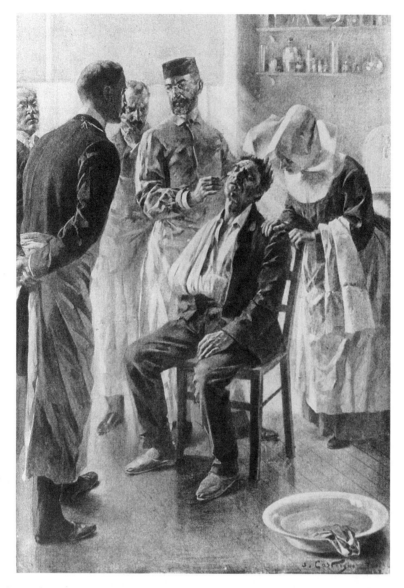

Figure 14. *A physician treating a lupus patient with radium at the St. Louis Hospital in Paris.*

nal cancers, but that the effects upon deep-seated cancers have not thus far proved satisfactory.

It has occurred to me that one reason for the unsatisfactory nature of these latter experiments arises from the fact that the rays have been applied externally, thus having to pass through healthy tissues of various depths in order to reach the cancerous matter.

The Crookes' tube, from which the Roentgen rays are emitted, is of course too bulky to be admitted into the middle of a mass of cancer, but there is no reason why a tiny fragment of radium sealed upon a fine glass tube should not be inserted into the very heart of a cancer, thus acting directly upon the diseased material. Would it not be worth while making experiments along this line?[4]

Bell's letter and the positive reply from the physician were published in the August issue of *American Medicine*.

Following Bell's idea, doctors began placing tubes, and later needles, containing radium directly into cancerous tumors. It was found that the element shrank the tumors, and in some instances completely destroyed the malignancy, thus curing the patient.

Internal Use

Radium also was used internally to treat hundreds of diseases, including everything from acne to insanity. It was administered orally, by inhalation and injection, and by enema and suppository.

One of the first to suggest radium be used internally was the electrical engineer and former Edison assistant, William J. Hammer. Hammer, who obtained samples of radium in France, used the element to successfully treat a tumor on his own hand. In 1902, he proposed several methods of applying radium internally. He wrote:

> ... I formed the original conception of treatment of disease internally by three distinct methods, all of which I have worked upon during the past year and more: the first is the making of various solutions radioactive by placing them in contact with or in proximity with radium; generally by immersing the sealed tubes in these solutions. These solutions were intended for and were used to destroy microorganisms producing malignant diseases internally and which could not be reached by the radium directly. The second idea was the making of medicines in either liquid, powdered or pill form, radioactive in order that this imparted radioactivity or this new characteristic imparted to the medicine might act upon the stomach, intestines, and other internal organs. The third idea was to introduce radioactive medicines or liquids or both, into the system by means of cataphoresis; in other words, by means of an electric current carrying the properties of this medicine into the blood or into the underlying tissues where the local treatment was desired. Over four months ago I supplied the Flower Hospital with radioactive solutions which they have used successfully and have continued to experiment with ever since. Prior to this time, and as far back as last Spring, I made other solutions, and used them upon myself and others.[5]

In 1903, a new discovery would greatly contribute to the internal use of radium. That year, the English physicist and discoverer of the electron, J. J. Thomson (1856–1940), found that well water in Cambridge, England, was weakly radioactive.[6] Others soon detected radioactivity in the springs of Europe and the United States. Some of the springs were found to contain small amounts of radium salts, while others contained radon, the radioactive gas given off by radium as it decays.

The presence of radioactivity in these waters seemed to explain the long held belief that some of the famous natural springs had remarkable healing powers. It was well known, for example, that for thousands of years people had traveled to the springs of Bath, England; Bad Gastein, Austria; Vichy, France; and St. Joachimsthal, Bohemia, to drink and bathe in their waters to cure numerous ailments.

In the United States there also were a number of well-known health springs, the most famous of which was Hot Springs, Arkansas. These springs were considered so important medically that the U.S. Government, by Act of Congress, made the area the first national reserve in 1832, setting it aside for the perpetual use of the American people. Later, Congress established a large military hospital at the springs to make the therapeutic benefits of the waters available to soldiers and sailors.[7]

Now that science had provided an explanation for the curative properties of the springs, people began to flock to them. Existing health resorts and spas expanded, and new ones sprang up. They also began advertising and promoting their radioactive waters. One spa at Hot Springs, for example, publicized that it had "wonderful curative results," because its "hot water [was] cooled without exposure to air," thus saving "all radium gasses ... to benefit [the] bathers."[8]

Many of the resorts and spas also began bottling their waters for their guests to take home with them. However, it was soon found that the vital curative radium gas (radon) in the waters did not last, but quickly escaped into the air.

In 1912, a poor California invalid who had studied the uses of radium would provide a solution to the problem. That year, R. W. Thomas patented the Radium Ore Revigator. Thomas wanted to create a device that would enable the masses to experience the health benefits of radioactive water in the comfort of their own homes and offices.

To create such a device, Thomas purchased some carnotite ore from Colorado for a few dollars a pound. He then mixed the radium-bearing ore with sawdust and molded it into an earthen jar, which he fired. Water placed in the jar would then accumulate radon as the radium decayed (Figure 15).

Thomas loaned some of the jars to sufferers of various chronic diseases. And they supposedly obtained remarkable relief from such conditions as constipation, rheumatism, high blood pressure, goiter, gout, malaria, stomach trouble, kidney problems, and many other ailments, "just like the relief afforded by the famous springs themselves."[9]

The reputation of Thomas's "magic jar" or "jar of life" quickly spread and he went into business to meet the growing demand, forming the Radium Ore Revigator Company of San Francisco. The company also established branch offices in several other California cities as well as in Chicago and New York. From 1915 until 1935, the company sold more than 100,000 of the Revigators.[10]

Figure 15. Revigator water jar (missing lid), ca. 1920.

Seeing the success, many other companies also began producing radioactive water jars for the public. For example, the National Radium Corporation of Chicago and Denver manufactured a metal jar called the Lifetime Radium Vitalizer (Figure 16). This device, which was made of aluminum, contained carnotite ore, which was held in the bottom by a perforated metal plate.

Another company that produced a radioactive water jar was the Radium Life Corporation of Los Angeles, which made the Radium Emanator (Figure 17). This unique device consisted of a metal container that held as many as five stackable plates made of concrete and carnotite ore. The plates' large surface area guaranteed enhanced emission of radon, and as the cure progressed, the plates could be removed to lower the amount of radon in the water.[11]

Many of the rich and famous of the time drank radium water, including industrialists, socialites, and politicians. For example, the mayors of New York City and Chicago, U.S. congressman and senators, and even the President of the United States, Franklin D. Roosevelt (1882–1945), supposedly drank it.

Although drinking radium water provided some radioactivity, the dose was generally very small and the effect was of short duration. Physicians, however, soon found a far better way to bathe the organs of the body with the miracle element by directly injecting it.

Figure 16. Lifetime Radium Vitalizer, ca. 1926.

Some of the first to experiment with radium injections were Austrian and German physicians. They found that radium appeared to be a remarkably effective internal medicine, and many of them enthusiastically reported their findings in numerous medical journals.

In 1913, two physicians from Johns Hopkins University, Drs. Rowntree and Baetjer, reviewed the extensive European literature on the internal use of radium. They found that of the 1,038 reported patients given radium treatments, a remarkable 837, or over 80%, were said to have benefited. Publishing their review in the *Journal of the American Medical Association*, the two physicians concluded:

Figure 17. Radium Emanator, ca. 1925–1935.

The value of radium is unquestionably established in chronic and subacute arthritis of all kinds ... acute, subacute and chronic joint and muscular rheumatism ... in gout, sciatica, neuralgia, polyneuritis, lumbago and the lancinating pain of tabes. In certain other conditions it is said to be of some value, although more data are necessary before this can be accepted—chronic bronchitis, chronic pharyngitis, pneumonia, myocarditis, arteriosclerosis, vasomotor disturbances, Raynaud's disease, scleroderma, idiopathic enlargement of the lymph-nodes and in chronic constipation.[12]

The internal use of radium unquestionably had physiological effects. It lowered blood pressure, relieved all types of pain, and stimulated the body's blood-producing system. There also appeared to be little risk in using it: there were few adverse side effects, and most of the element was quickly eliminated from the body through the feces.

No one realized at the time that some of the radium might be deposited in the bones and retained there throughout the person's entire natural life, and traces of it would be found even long after death, for thousands of years. They also failed to comprehend that years and even decades later, the element would damage and destroy the body's blood-producing system, and cause severe anemia and bone necrosis, as well as various types of cancers.

There were, in fact, warnings concerning the internal use of radium. For example, in May of 1912, a German physician in a Berlin hospital reported the death of a patient from radium. He had treated a 58-year-old woman who was suffering from chronic arthritis with frequent injections of thorium X (a short-lived isotope of radium derived from mesothorium) for a period of 16 days. The woman began to suffer from diarrhea and severe bleeding, and she died within a month.[13] The attending pathologist reported "… one cannot doubt for a moment that we have here a case of death caused by mesothorium."[14] This incident, as well as other warnings, however, were totally ignored.

In the United States, the first physician to experiment with radium injections was Dr. Frederick Proescher of the Standard Chemical Company of Pittsburgh. The radium company had established a laboratory to investigate the possible medical uses of the element. And to direct the new facility, the firm had recruited Proescher from Germany. In 1913, Proescher began injecting a few patients with small doses of radium. He soon found the results so encouraging that he began treating more patients, giving them larger doses of the element.[15]

The results of Dr. Proescher's experiments seemed so promising that the Standard Chemical Company established a free radium clinic in Pittsburgh. Later, other radium clinics were opened in New York City and Chicago.

In 1914, the influential American Medical Association tacitly approved the internal use of radium by adding it to its list of "New and Nonofficial Medical Remedies." The association also began developing standards and testing medicines and various health products to assure that they really contained the element, and that they were sufficiently radioactive (Figure 18).

Over the next several years, the use of radium injections greatly increased. One of the most frequent uses of the injections was to treat arthritis. It became such a common practice that many physicians did not even bother telling their patients they were receiving the element.[16] A few unscrupulous physicians, however, indiscriminately gave their patients a large number of the injections for almost any condition.

One particularly overzealous physician and charlatan was Dr. C. Everett Field (1870–1951) of New York. After graduating from medical school, Field had worked for the Standard Chemical Company's radium clinic in Pittsburgh, where he learned how to

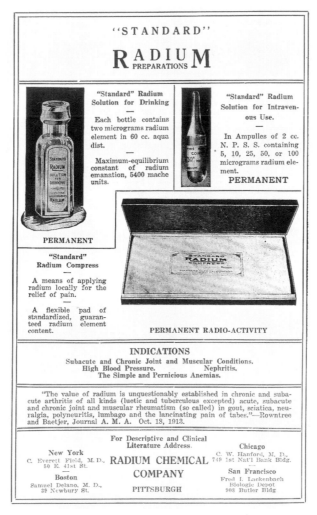

Figure 18. An advertisement from the journal Radium, *1914.*

use the element. A few years later, he founded the Radium Institute of New York City. At the institute, he treated thousands of patients with radium injections. In 1921, for example, Field wrote that he had administered over 7,000 radium injections in eight years.[17]

An article in a 1920 issue of *Scientific American* summed up the view of many physicians of the time regarding the internal use of radium. It declared:

> ... radium may even be taken internally for the treatment of rheumatism, certain kidney and liver troubles, and intestinal troubles. Of course only a minute amount is taken, usually in a liquid. The radium is absorbed into the blood and circulates throughout the system. No actual chemical action takes place, however, even when the radium is taken internally. It simply attacks useless tissues, which are then ejected from the body by the excretory system.[18]

Chapter 4
Mysterious Deaths

> *The history of this particular patient is so evident, that there is practically no doubt in my mind as to the responsibility of the U.S. Radium Corporation. ... Unfortunately we have not been able to find out so far what chemicals they are using for painting the numerals on watch dials.*[1]
> Dr. Theodor Blum, 1924

Radium Paint, Watches, and Two Physicians

The first person to invent radium-luminous paint was the American electrical engineer William J. Hammer. In late 1902, he mixed zinc sulfide with radium he purchased in France to produce the world's first radium paint. He applied this new substance to numerous items, including watch and clock dials. Hammer, however, as previously noted, did not attempt to obtain a patent for the paint, because radium was so scarce and costly.[2]

Several months later, the famous American gem expert George F. Kunz (1856–1932), who was aware of Hammer's work, also created a radium-luminous paint. Kunz, a vice president at the jewelry firm Tiffany & Company of New York City, applied the paint to several jewelry items and to a watch dial. In September of 1903, Kunz filed a patent for his paint, or "cold light," as he called it.[3]

Although radium paint was first invented and patented in America, little was done with it, and for many years it remained only a novelty. In Europe, however, radium paint was swiftly put to commercial use. Around 1905, it was being applied to the dials of expensive Swiss watches and clocks. In a short time, thousands of small Swiss shops were using the glowing paint. There were so many radium painters in that country that it was common to recognize them on the streets, even on the darkest nights, because of the glow around them: their hair sparkled almost like a halo.[4]

Over the next several years, many other radium dial-painting firms also were established in France, Germany, and England. It was not until 1913, however, that the first American firm commercially produced radium-luminous paint. That year, a small laboratory manufactured and applied it to several thousand watch dials. In time, the laboratory would become the U.S. Radium Corporation, producing millions of radium dials and becoming the single largest radium application firm in the world.

The corporation was founded by two physicians: Sabin A. von Sochocky (Figure 19) and George S. Willis. Despite their knowledge of the dangers of radium, both doctors repeatedly ignored the risks and carelessly used the new radioactive element. Both would eventually suffer horribly for their recklessness.

Dr. Sabin Arnold von Sochocky (1883–1928) was born in a part of the Austrian-Hungarian Empire that is now in Ukraine. He studied physics, chemistry, and medicine at the University of Lvov. Later, he went on to the University of Moscow, where he received his medical degree. Von Sochocky also took various courses at the universities and technical institutes of Vienna, Prague, and Dresden. While traveling in Paris, he met the Curies and became interested in radium.[5]

Dr. von Sochocky came to America in 1906, and practiced medicine for a number of years. Around 1913, he began experimenting with radium, establishing a small laboratory in New York City. Although he was interested in using radium for medical purposes, he first created a radium-luminous paint, which he used to finance his medical research, by applying it to watch and clock dials.[6]

Von Sochocky's radium paint soon came into great demand. His laboratory received a large contract to paint the dials of inexpensive dollar watches produced by the Ingersoll Watch Company (Figure 20). Contracts from other watch and clock companies soon followed. Von Sochocky's laboratory quickly became too small, and a much larger building had to be found where factory-scale production could be used.[7]

In 1914, Dr. von Sochocky closed his laboratory, and along with Dr. George S. Willis, a prominent New Jersey physician, organized the Radium Luminous Material Corporation. They located the new corporation in Newark, New Jersey. During the next few years the firm's dial-painting operations continued to expand.[8]

When America entered World War I in April 1917, it created an enormous military demand for many types of radium-treated devices, and for workers to paint them. Radium dials were needed for the instruments on airplanes, warships, and submarines, and for soldiers' compasses and watches. To meet the demand, the company greatly expanded, hiring hundreds of new workers. At its peak, it employed 250 young women to paint radium dials.[9]

That same year, the corporation needed so much radium that its management decided to produce the element itself. The firm acquired carnotite mines in Colorado and moved all of its operations from Newark to nearby Orange, New Jersey, where the corporation established a radium extraction plant, laboratory, and dial-painting studio (Figure 21).[10]

In 1918, an estimated 95% of all of the radium produced in America was used to manufacture radium paint for military purposes.[11] At the front, radium paint was not only used on dials but also to mark safe passage through barbed wire zones. The paint was used on the backs of trucks so they could be seen at night.[12] And radium paint was

Figure 19. Dr. Sabin Arnold von Sochocky.

used for the gunsights of machine guns and artillery. By the armistice in November 1918, practically every American soldier wore a radium dial watch.[13]

After the war, consumer demand for radium watches skyrocketed. The public was fascinated that such a mysterious, scarce, and costly element as radium could be used on inexpensive watches and clocks, and that virtually everyone could afford to own a timepiece with the magical element on it. In 1919, 2.2 million radium dial watches and clocks were produced in America.[14] A year later, the number had increased to 4 million.[15]

At the same time, radium paint also was being applied to a wide variety of consumer products including flashlights, pull-chain pendants, push-button switches, door bells, house numbers, door locks, safe combinations, automobile and motorcycle gasoline gauges

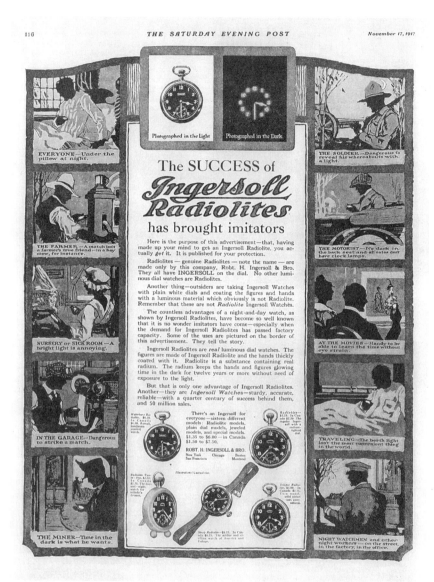

Figure 20. Advertisement for Ingersoll Radiolite watches from the Saturday Evening Post, *1917.*

and speedometers, telephone mouthpieces, convention buttons, buttons on bedroom slippers, poison bottle indicators, fishing lures, and the eyes of toy animals and dolls.[16]

In 1919, the Radium Luminous Material Corporation made several important changes. It began using mesothorium, a cheaper isotope of radium. The company also began selling radium paint directly to its customers for application at their plants. Instead of sending watches to Orange, New Jersey, to have their dials painted, the company would now train the workers at the watch factories to paint dials. That year, the company assisted one of its largest customers, the Waterbury Clock Company of Waterbury, Connecticut, to establish a dial-painting department.[17]

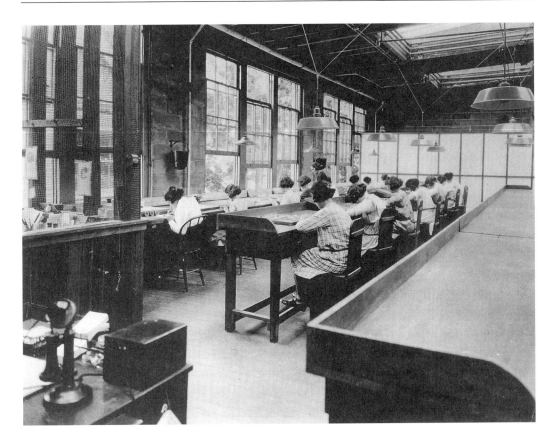

Figure 21. *Dial painters working at the U.S. Radium Corporation, ca. 1924.*

Also in 1919, the Radium Luminous Material Corporation's radium-processing operations produced more of the element than it needed. And in the following year, it began selling radium. To mark the change, in 1921 the company was renamed the U.S. Radium Corporation.[18]

Dr. von Sochocky, who was the corporation's president and technical director for several years, was fascinated with radium. Although he was well aware of the fact that radium could be deadly, he liked to play with the radioactive element. Sometimes at night, when he was working late in his laboratory, he would take tubes of radium out of the company safe and hold them in his bare hands and watch them glow. He also would immerse his bare arm up to the elbow in solutions of radium.[19]

In time, von Sochocky's carelessness would take its toll. He frequently burned his fingers from directly handling tubes of radium. Some of the burns would take many months to heal. Other burns were so severe that they destroyed the bones of his fingers. Von Sochocky, however, paid little attention to his injuries. When the tip of his left index finger turned black from radium necrosis, he simply hacked it off.[20] In discussing one radium burn that destroyed part of the bone in his thumb, he wrote:

> This...is nothing compared with the risks and sacrifices which scientific experimenters are undergoing all the time, and I would not mention it if I had not been asked to do so in telling you just as plainly as I can all about the remarkable powers of radium.[21]

Around 1920, von Sochocky began to suffer from transitory anemia. He treated himself, and temporarily moved to Colorado, hoping that the high altitude would keep his disease in check. In time, he began to suffer from radium necrosis of the upper and lower jaw, which resulted in the loss of his front teeth. His fingers up to the second knuckle were also black from the same condition.[22]

Von Sochocky remained in fair health until August 1928. That month, however, he became extremely weak, pale, and his breathing was labored. Day by day, his anemia worsened. He was hospitalized for three months, and given numerous blood transfusions, which helped for a while. Finally, his body stopped producing blood of its own. And on November 14, 1928, Dr. von Sochocky died at the age of 45 from radium-induced aplastic anemia.[23]

Dr. George S. Willis (1876–1924) also would experience the harmful effects of radium. Willis, who was born in Brooklyn, New York, obtained his medical degree from the New York Homeopathic Medical College and Hospital of New York City in 1899. After graduation, he began practicing medicine in Morristown, New Jersey.[24]

Around 1914, he and Dr. von Sochocky began experimenting with radium. That same year, they organized the Radium Luminous Material Corporation, with Willis serving as the chairman and director of the new corporation. From 1915 until 1920, Willis began working with large amounts of radium in his medical practice, but he took no precautions while handling tubes of radium, picking them up with his right thumb and forefinger.

In 1918, Willis began experiencing a curious numbness in the tips of his fingers. Soon his hands became rough. By early 1920, the skin of his fingers became tender and sore, and he felt a constant burning sensation. In time, fissures developed on his fingers and thumb, and his hand required constant medical care. The pain became so severe that he could not sleep at night, and he had to keep his hand elevated all the time. In September 1922, a fissure on his right thumb became extremely painful, and the digit had to be amputated. Dr. Willis had developed a radium-induced cancer of the thumb.[25]

In 1923, Willis co-authored an article in the *Journal of the American Medical Association* describing his experience. In the article, he warned others that:

> The reputation for harmlessness ... enjoyed by radium may after all depend on the fact that, so far, not very many persons have been exposed to large amounts of radium by daily handling over long periods. With the use of large quantities of radium there is good reason to fear that neglect of precautions may result in serious injury to the radium workers themselves.[26]

Unfortunately, the warning would be ignored, and it would come too late for many.

The Short, Tragic Life of Mollie

In October, 1917, a 20-year-old woman by the name of Amelia Maggia (1896–1922), or Mollie as her friends called her, began working as a dial painter for the Radium Luminous Material Corporation. Mollie's sister, Albina (1895–1946), who was a year older, had started working for the company several months earlier as a dial painter. And by the end of the year, Mollie's younger sister, Quinta (1900–1929), who was 16, also took a job as a dial painter.

Mollie's parents were Italian immigrants. Her father, Valerio, and mother, Antonetti, came to America, and settled in Orange, New Jersey. Valerio worked as a hatter, and Antonetti remained home to raise the couple's seven daughters.

Mollie and her two sisters thought they were lucky to be working together at the same plant. Further, the work seemed ideal: it was glamorous to apply the mysterious, glowing paint, and, depending upon the number of dials they painted, it could pay very well.

At the company, the sisters joined the many other dial painters that the firm had recently hired. All the painters were young women, generally between the ages of 16 and 20. Most of them would remain with the firm for a year or two; only a few would continue to work for five or six years, until they married and left to raise a family.

The sisters, like all of the other dial painters, were taught the tasks of preparing and applying the paint. They were shown how to mix a small quantity of fine yellow powder, which contained the radium, with an adhesive in a small crucible; how to frequently stir the material to keep it in suspension; and how to apply the paint to the numerals and the hour and minute hands of watches using the tip of a very fine, small brush.

The paint used by the company contained an infinitesimal amount of radium. In fact, there was only a little more than one-millionth of a gram in the paint applied to the dials of 50 inexpensive watches.[27] This amount was so small that it could not even be seen with the naked eye. For the dials of more expensive watches and instruments, more radium was used; however, it still was an extremely minute quantity.

To make the luminous paint, radium bromide was mixed with a large amount of zinc sulfide to form a fine yellow powder. The radiation that was emitted by the radium made the tiny zinc sulfide crystals in the powder flash, or scintillate, giving the material the appearance of continuously glowing.

Depending upon the type of application, either an oil adhesive or a water paste was added to the powder. When the paint was being applied to metal, ivory, or porcelain dials, such as those on expensive watches and instruments, an oil adhesive containing alcohol and turpentine was used. In applications to paper or cardboard dials, such as those on inexpensive dollar watches, a thin, water-soluble paste containing acacia (gum arabic) was employed.[28]

All of the dial painters worked on a piecework basis, receiving about eight cents for each dial and pair of hands they completed (the equivalent of about 70 cents today). Thus, the more they painted, the more money they earned. Like most workers of the time, they received no other benefits such as vacation time or sick days.

The instructors who taught the new dial painters how to apply the paint told them that although it was against company rules to tip or point the brush with their lips for sanitary reasons, it would speed the process, and it was the only way for them to make really good money. The company's management, which was well aware of the practice among its employees, never strenuously objected to it.[29]

Although some of the dial painters rarely tipped their brushes, others frequently did, pointing them each time they applied the paint to a numeral and hand. Most workers found the paint with the water paste to be palatable; although it was tasteless, it did have a gritty texture. In contrast, the paint with the oil adhesive was very distasteful, even nauseating, and rarely did a dial painter point the brush with it.[30]

The number of dials the women painted varied considerably. Some employees painted only 30 dials a day, while a few highly skilled workers could paint as many as 300.

During the war years of 1917–1918, there was such a tremendous demand for radium dial watches and military instruments that many of the dial painters worked seven days a week. Many believed they were performing a patriotic duty and significantly contributing to the war effort.

Sometime during a work break, some of the women would write their names and addresses on the inside of a watch they were painting, hoping that a handsome young soldier would find it and write back to them.

After the war, in 1919, Mollie's two sisters, Albina and Quinta, left the company to marry. Mollie, however, continued to work. And, later, she was joined by another one of her sisters, Irma (1903–1940), who also was hired as a dial painter.

Around 1921, Mollie began having trouble with her teeth. She developed a toothache, and went to a dentist to have it extracted. At about the same time, she noticed vague, rheumatic-like pains in various parts of her body. And she became anemic, pale, and lost weight.

In October, Mollie again experienced problems with her teeth. She went to see Dr. Joseph P. Knef (1879–1946), a Newark dentist. After examining her mouth, Knef found that the socket of the tooth that had previously been removed had failed to heal, so he began to treat her.

Mollie's health continued to decline and in January of 1922, she went to see a physician. The doctor was puzzled by her condition. He administered several medical tests, including a Wassermann test for syphilis, which proved negative. That same month, because of ill health, she quit her job after working for the company for more than four years.

In March, Mollie visited another physician. She told the doctor about the aches and pains from which she suffered, particularly in her joints. The doctor diagnosed the cause as rheumatism, which he treated with aspirin.

Mollie's condition continued to worsen. The pain in her teeth and jaws was becoming unbearable. Again, she went to see Dr. Knef, who found that her jaw was slowly disintegrating. To stop the necrosis, he resorted to extreme methods of treatment; however, the more he did, the faster the disease spread. After seven months of treatment,

Mollie's jaw bone had so disintegrated that Knef simply removed pieces of it by lifting it out with his fingers. A week later, her entire lower jaw was removed in a similar way.[31]

In June, Dr. Knef also gave Mollie a Wassermann test. The laboratory analyzing the test results, however, misdiagnosed her as having syphilis. When Knef received the test result he told her physician, who gave her a series of injections for the disease.

During the next few months, Mollie's health declined day by day. She suffered from severe anemia. The necrosis continued to spread, to the roof of her mouth and even to the bones in her ears. And she began to increasingly bleed from the mouth.

On September 12, 1922, at the age of 25, Mollie died. Although later she would be recognized as the first known death from radium poisoning, the cause of death would be erroneously listed on her death certificate as due to "ulcerative stomatitis, syphilis."[32] This verdict compounded Mollie's tragedy; not only did she die an agonizingly painful death, but the stigma of supposedly having venereal disease added to the trauma and loss suffered by Mollie and her family.

Mollie's tragic death would be the first of many. Years later, fatal radium poisoning would claim her three other sisters who also had worked at the plant. Two of them would file a lawsuit against the company; in the most dramatic event of the resulting spectacular trial, Mollie's body would be exhumed to be tested for the presence of radium. Every portion of her tissue and all of her bones would be found to contain the radioactive element.[33]

Speculation and False Starts

Although Mollie's mysterious illness and death did not set off alarms in the medical community or radium industry, the illness of a second dial painter would lead to several investigations. Irene Rudolph (1902–1923) was born in Newark. Both of her parents died when she was young, and she lived with her cousin. At the age of 16, Irene started working as a dial painter for the Radium Luminous Material Corporation. She worked for the company for two and a half years, from September 1918 until April 1921.

In the spring of 1922, Irene's cheek started to swell. She visited a dentist, who extracted a tooth and thought she was cured. However, the area refused to heal, and her teeth and jaws slowly began to disintegrate.

In July, Irene was referred to Dr. Walter F. Barry (1878–1942), a distinguished Newark oral surgeon.[34] To stop her necrosis, Dr. Barry and his partner Dr. James B. Davidson (1887–1957) treated her, removing her lower teeth and part of her jaw bone. The disease, however, continued to spread.

Both Barry and Davidson believed that Irene was suffering from the horrible occupational disease phosphorus necrosis or "phossy jaw." The disease, which had been all but eliminated from the nation's match industry, had recently been found in New Jersey among workers who manufactured fireworks.[35]

The two dentists were so troubled by Irene's condition that they went to the radium plant and confronted two of its managers. They accused them of poisoning her and the other dial painters by using white phosphorus in their paint. Dr. Davidson told one of the managers that "he ought to close down the plant; they had made $5 million dollars; and why go on killing people for more money."[36] The managers, however, vigorously denied any responsibility. And when they told the dentists that the company's paint had never contained any phosphorus, Dr. Barry told them that it must be the radium, or its radioactive constituents that caused Irene's condition.[37] After the confrontation, the managers immediately instructed the company's dial painters to stop putting their brushes into their mouths.

Irene's condition continued to worsen, and Dr. Barry referred her to a surgeon. In December, she was admitted to a Newark hospital, where medical tests revealed that she was suffering from osteomyelitis (inflammation of the bone marrow) of the entire jaw. Pus was found to be extruding from the roots of her molars. And she also was found to be anemic.

Irene stayed at the hospital for several weeks, where she was seen by Dr. George Herbert Allen (1895–1943), a physician and anesthetist. Dr. Allen also believed that she was suffering from the occupational disease of phosphorus poisoning. On December 26, 1922, he notified the Newark Public Health Department of her condition.[38]

As soon as the director of the health department, Dr. Charles V. Craster (1878–1953), heard of the case, he immediately began an investigation. He was very concerned that if it were a case of the occupational disease, it would only be a matter of time before more of the dial painters developed it.

Dr. Craster sent a health inspector from the department's industrial hygiene division to the hospital where Irene was to obtain detailed information concerning her medical and work history. He also sent an inspector to the radium plant to call their attention to the fact that the dial painters were pointing their brushes in their mouths.

When the inspector told the company's vice president and production manager about the situation, the manager assured him that he had repeatedly warned the women against this dangerous practice, but he could not get them to stop it.

Because the radium company was not located in his jurisdiction, Dr. Craster wrote to the deputy commissioner of the New Jersey Department of Labor, John Roach, and asked him to conduct a health and safety inspection of the plant. Craster also forwarded a sample of the radium paint used by the company to the labor department, asking them to have it chemically analyzed to see if it contained any white phosphorus.[39]

The labor department quickly sent an inspector to the radium plant. She investigated the plant's radium recovery and dial-painting methods. At the end of January 1923, she wrote to Roach indicating that she was convinced that the luminous paint used at the plant did not contain phosphorus. However, she also erroneously wrote that the paint did not contain any radium, but was made radioactive by some secret process involving the element.[40]

The labor department sent a sample of the paint to its consulting chemist for analysis. Even before starting the task, the chemist wrote to Roach expressing his opinion, stating: "As you know, radium has a very violent action on the skin, and it is my belief that the serious condition of the jaw has been caused by the influence of radium."[41]

When the paint was analyzed, it was found not to contain any phosphorus whatsoever. The chemist notified Roach of the fact, however, he again confirmed his belief that radium was responsible for the problem. He wrote: "... I feel quite sure that the opinion expressed in my former letter ... is correct, and that such trouble as may have been caused is due to the radium contained therein."[42]

The labor department continued its investigation, conducting several other inspections of the plant. No violations of state law, however, were ever found, and the plant continued to produce radium and paint watch dials. According to one author:

> The State Labor Department, accustomed to conventional inspection as to hours of work and number of windows for ventilation, gave the plant a clean bill of health. Even after the fatalities were known and the presence of excessive radium dust authenticated, the Department stated that they had no legal authority to shut down any process in a factory.[43]

Over the next several months, Irene Rudolph grew sicker. In July of 1923, she again entered the hospital, and a day later, she died at the age of 21. Her cause of death was erroneously listed as "phosphorus poisoning, resulting in the complete necrosis of the lower jaw."[44]

At the same time, other radium dial painters also were getting sick, including Hazel Vincent Kuser (1899–1924). Hazel was employed as a dial painter from 1917 until 1920. She continued to work until her illness forced her to leave the company.

While she was working, she began to have dental problems. Following the pattern set by other radium dial painters, she went to a local dentist for care. In September of 1923, she had a tooth extracted. And she soon developed necrosis of the jaw, which necessitated removal of much of her upper jaw and required two blood transfusions. Hazel went to a young oral surgeon for relief. However, he was so puzzled by her condition that he referred her to Dr. Theodor Blum (Figure 22).

Dr. Blum (1883–1962) was a prominent oral surgeon and clinical oral pathologist. He was one of the nation's first oral surgeons. He also was one of the first to use x-rays for dental diagnosis and novocaine as a local anesthesia. Born in Vienna, Austria, Blum came to America in 1904. He studied dentistry and medicine at the University of Pennsylvania where he earned doctoral degrees in both. In addition, he briefly returned to Austria where he earned another doctorate in medicine from the University of Vienna. In 1912, he established his dental practice in New York City.[45]

In January of 1924, Dr. Blum saw Hazel for the first time. When he examined her mouth, he quickly realized that he had never seen a condition like hers before. She seemed to be suffering from a strange kind of malignancy of the bone, which had no source and no known cause. He asked her about her medical history and past employment. After Blum learned that she had worked at a radium plant and had pointed her brush in her mouth, his diagnosis was that she suffered from osteomyelitis of the jaws and poisoning from a radioactive substance.[46]

Figure 22. Dr. Theodor Blum.

After her visit, Dr. Blum had Hazel admitted to a hospital where he operated on her jaw. She recovered and was discharged. But, her condition continued to worsen and she was hospitalized several more times.

Despite her deteriorating health, Hazel's boyfriend, Theodore Kuser, who had been her childhood sweetheart, insisted that they marry, so he could take care of her. To pay for her mounting medical bills, her new husband mortgaged everything he owned, and he borrowed money from his family. The expenses, however, rose so rapidly that he and his family quickly became destitute.

In June 1924, Blum was so concerned about the couple's financial situation that he wrote to the radium corporation requesting that they help pay for Mrs. Kuser's medical bills. The corporation, however, promptly wrote back, refusing to help in any way, stating that it would establish "a precedent which we do not consider wise."[47]

Over the next months, Hazel suffered excruciating pain. In December 1924, after suffering for more than four years, she died at the age of 25. The cause of death was listed on her death certificate as occupational poisoning, necrosis of the jaw and maxilla, and profound anemia. The certificate, however, also indicated that the "deceased was employed painting dials on radium-faced watches, using a fine brush—at times moistening edge with mouth."[48]

Before Mrs. Kuser's death, Dr. Blum added a brief footnote describing her condition to a paper that he had previously delivered at a dental society meeting. The paper was scheduled to be published in the September 1924 issue of the *Journal of the American Dental Association*.

In it he wrote:

> During the Fall of 1923, there came under my observation a case of osteomyelitis of the mandible and maxilla, somewhat similar to phosphorus necrosis; which, however, was caused by some radioactive substance used in the manufacture of luminous dials for watches. The condition has been termed 'radium jaw.'[49]

This footnote would be the first reference to the new occupational disease in the world's medical literature. Although Dr. Blum believed for the rest of his life that his note led other researchers to investigate the new condition, the fact was it went completely unnoticed.[50] It would take several other investigations to conclusively prove the existence of the mysterious, deadly new disease.

Chapter 5
Medical Detectives and Social Activists

To a lay person, it seems impossible that a "coincidence" can account for the fact that four persons have died, another is dying and others have apparently the beginning of the same trouble, when all of them have no common experience other than the same occupation.[1]

    Katherine G. T. Wiley, 1925

The Harvard Public Health Team

As more of the dial painters sickened and slowly began dying, Arthur Roeder, the president of the U.S. Radium Corporation, became increasingly concerned. Rumors were spreading that the company's paint was deadly. Workers at the radium plant were suspicious and fearful, and the corporation was finding it increasingly difficult to hire new employees. Nevertheless, Roeder felt optimistic that he could solve all these problems.

Arthur Roeder (1884–1960), the son of a minister, was a highly successful businessman. Born in New Jersey, he studied civil engineering at Cornell University for four years, but left without receiving a degree. Returning to New Jersey, he took a job as a junior engineer with the City of Newark. In 1910, he joined the Robert Ingersoll and Brother Company of New York City, the world's largest watch manufacturer. At the firm, Roeder quickly became a success. In only eight years, he rose from salesman to assistant sales manager, then to production manager, and finally to assistant to Charles Ingersoll, the general manager. In 1918, however, Roeder left the watch company to become the treasurer of the Radium Luminous Material Corporation. And when the company was reorganized and renamed the U.S. Radium Corporation in 1921, he was appointed president (Figure 23).[2]

In early 1924, when Roeder was faced with the New Jersey labor department's investigation of the radium plant, he decided to have all of the company's current dial painters examined by the Life Extension Institute of New York City. The institute was a highly reputable organization devoted exclusively to administering periodic physical examinations to workers for preventive health care purposes.[3]

Figure 23. *Arthur Roeder.*

At the same time, Roeder contacted a friend who was a general manager at the New Jersey Zinc Company. Roeder thought that perhaps the zinc sulfide contained in the company's radium paint might be the cause of the health problems. This friend, however, reported that the zinc company had not experienced any problems whatsoever with the material, but did recommend that Roeder contact Dr. Cecil K. Drinker of Harvard University, who would undoubtedly be interested in investigating the situation.[4]

After examining the dial painters, the Life Extension Institute reported its findings to the company. The institute found that nothing was wrong with their health. The next day, March 12, 1924, Roeder wrote to Dr. Drinker, requesting that he conduct an investigation. In his letter, he explained that two former workers had developed jaw necrosis, and that one had died, while the other was supposedly recovering.[5]

Figure 24. Dr. Cecil K. Drinker.

Dr. Cecil Kent Drinker (1887–1956), a member of an old and distinguished Quaker family, was a physician and professor of physiology at the Harvard School of Public Health, and a renowned expert on industrial hygiene and occupational medicine (Figure 24). Drinker also was one of the first experts to stress the importance of the respiratory tract as the route of absorption of toxic dust and fumes. And together with his brother, Philip, who also was a professor at the Harvard School of Public Health and one of the first bioengineers, they developed ventilation methods and respiratory devices for protecting workers in the dusty trades. Philip Drinker would later gain world fame as the inventor of the Drinker Respirator, or "iron lung," which was widely used in treating victims of polio.[6]

On March 15, Dr. Cecil Drinker wrote back to Roeder, accepting his request. Drinker indicated that the jaw necrosis of the two workers might prove to be nothing more than a coincidence: however, he felt in order to be safe, a complete investigation was necessary. He agreed to visit the radium plant, to see the dial painters at work, and to talk to those who had medically examined the women.[7]

A month later, Dr. Drinker and his physician wife, Dr. Katherine R. Drinker (1889–1956), visited the Orange, New Jersey, factory. During their one-day trip, they briefly toured the application plant. They saw the workers applying the radium paint, and they learned that six months earlier, the dial painters had stopped pointing their brushes in their mouth. The company provided the couple with the Life Extension Institute report. And they also were taken to see the two dentists, Drs. Davidson and Barry, who had treated one of the dial painters and who had previously confronted the company's management, accusing them of poisoning their workers.

Only two weeks after their visit, Dr. Cecil Drinker wrote back to the corporation indicating:

> From material which we have been able to dig out of the literature here and piece together with your experience ... it would seem that radium is the probable cause of the trouble. ... There seem to be two possibilities in regard to the radium: first, that the rays are causing the damage; and, secondly, that radium itself, absorbed in minute quantities through the skin over long periods of time, is deposited in the bones. Since it apparently behaves like calcium this point of deposition seems highly probable to us. Once deposited in the bones, my associates who have been working with radium feel that it might exist for a good while and continue to slowly exert harm.[8]

On May 7–8, Drinker, along with his wife and Dr. William B. Castle (1897–1990), a young physician from the Harvard School of Public Health, conducted a two-day inspection of the radium plant. The team planned to study the factory carefully and to examine its employees medically.

The public health team quickly was appalled at what they found. The plant literally was saturated with radium-contaminated dust. It glowed everywhere: on the floors, chairs, workbenches, walls, light fixtures, and rafters.

When the dial painters were examined, they were found to have the sparkling dust in their hair, and on their faces, hands, arms, and necks. Even their underclothes and corsets shone. One dial painter had luminous spots on her legs and thighs, while another's back was luminous almost to her waist. One woman even reported that when she blew her nose, the discharge glowed.

To further determine the presence of atomic radiation in the factory, the team placed sealed dental film at various locations, and also had workers carry it for several days. After a short time all of the film fogged, indicating excessive exposure to gamma radiation, similar to x-rays.

Last, the team took blood samples from 22 workers, including one employee who had only worked at the plant for two weeks. Not one of the samples was found to be normal. They also found that those workers with the highest exposure to radium had the worst blood abnormalities.[9]

The radium company appeared to have an utter lack of realization of the dangers inherent in the radioactive material that it processed and used. The firm had not even followed such basic safety procedures as issuing protective clothing such as smocks to the dial painters.[10]

On June 3, 1924, Dr. Cecil Drinker sent the president of the corporation the team's final report. In its cover letter, he emphatically wrote:

> We believe that the trouble which has occurred is due to radium. We realize that no final proof of this can be offered at the present time, but in view of material in the literature and of facts disclosed by our investigation it would, in our opinion, be unjustifiable for you to deal with the situation through any other method of attack.[11]

Arthur Roeder, however, refused to accept the report's findings and safety recommendations. He claimed that the report was only preliminary, with tentative conclusions, and based on circumstantial evidence. Nevertheless, Roeder did have a notice posted in the plant forbidding the dial painters to put their brushes in their mouths. Later, he would falsely claim that the report completely cleared the company and state that it said that all of the dial painters were found to be in perfect health.

After receiving the Harvard report, Roeder quickly wrote back to Drinker, denying that radium could possibly be the cause of the problem. Roeder claimed that the firm's vice president had just compiled a voluminous internal report containing new evidence, completely nullifying the team's report. Over the next several months, Drinker repeatedly asked to see a copy of the alleged document. However, Roeder never sent it.[12]

At the same time, Roeder wrote Drinker, arguing that the women who were sick were former employees who had not worked for the company in many years. He stated that employees who "really" were exposed to radium (such as the company's chemists and radium processors) had no health problems. He insisted that the illness might have been caused by some unusual contagious infection outside or within the factory. Last, he pointed out that no other radium application plant in the nation, or in Europe, had experienced the mysterious disease.[13]

Drinker carefully and patiently wrote back, addressing each of Roeder's arguments. He knew that the president of the corporation was facing a dilemma: radium was just too profitable to admit that it could be deadly.

Drinker summed up Roeder's situation, writing:

> ... the unfortunate economic situation in which he finds himself makes it very hard for him to take any stand save one in regard to radium, namely that it is a harmless, beneficent substance which we all ought to have around as much as possible.[14]

Months later, when Drinker asked Roeder for permission to publish the Harvard team's report, he refused. The president of the U.S. Radium Corporation never wanted the incriminating report released. According to one author:

Obviously, this report by the Harvard investigators was a scientific document of the greatest importance, not only to remedy conditions in this plant, but to acquaint other manufacturers, using the same radium formula, with its toxicity and potentially lethal effects. Science and humanity alike demanded immediate publication of this report. But the U.S. Radium Corporation, having commissioned and paid for the investigation, now refused permission to publish. The facts, naturally, reflected on the plant; they might well serve as a basis for damage suits. Rumors of the investigation had leaked out, but the report was resolutely suppressed.[15]

Dr. Drinker, at first, reluctantly accepted Roeder's decision. Later, however, circumstances would compel him to publish the report anyway, over the corporation's strenuous objections and the imminent threat of a lawsuit.

The Consumers' League

In March of 1924, as the radium corporation was hiring Drinker, the New Jersey labor department requested the public health officer of the City of Orange, New Jersey, to help in its investigation of the radium plant, but not having enough time or resources, she declined. The health officer did, though, write to the Consumers' League of New Jersey, asking them to investigate the situation.

The New Jersey league was a chapter of the National Consumers' League, which was an important lobbying group for social and labor reform. The organization was particularly interested in improving the working conditions of women and children.[16]

The executive secretary of the New Jersey league, Katherine G. T. Wiley, quickly agreed to help. After obtaining a list of the names and addresses of dial painters who had died and who were ill from the mysterious disease, Wiley began her investigation. First, she went to the families of those women who had died. From them, she obtained information about the victim's past work experience, and a detailed history of the women's fatal illnesses.[17]

Wiley was immediately struck by the similarity in the details of the disease: the initial problems the women had experienced with their teeth and jaws; the difficulty in healing after a tooth was pulled; the horrific jaw problems that developed; and the fact that all the victims developed severe anemia.

Next, Wiley visited the women who were sick. She was shocked by their appearance and moved by their terrible suffering. The once vibrant young women were now like walking ghosts. Many had given up all hope of recovery. And they and their families had become destitute paying their growing medical and hospitals bills. Wiley vowed to leave nothing undone to find the cause of the disease, and once that was accomplished, to stick to the matter until something was done to assure that no more women would endure such horrible suffering.[18]

To obtain additional information, Wiley contacted all of the physicians and dentists who had treated the women. And she personally visited Drs. Barry, Davidson, Knef, and Blum. They all stated that the dial painters were suffering the effects of some kind of poison. Drs. Barry and Davidson went so far as to say that the mouths of the dial painters were radioactive. Wiley also learned that Dr. Drinker was conducting an investigation of the mysterious disease for the radium company.[19]

Wiley went to the New Jersey Department of Labor to discuss the situation with the commissioner and the deputy commissioner of labor. She was surprised to find that both of them already knew the data that she had collected. However, they indicated that despite those facts, they had no conclusive evidence that the cause of the deaths and illnesses were in any way connected with the women's occupation. They also stated that under state law they had no authority to stop an industrial process even if it was known to be injurious.[20]

Finding the labor department of no help, Wiley contacted a local judge as well as a lawyer from the Legal Aid Society of Newark. She found that the sick women were not eligible under New Jersey's limited workers' compensation program. Wiley also asked about the possibility of the victims filing a lawsuit against the radium company. The judge and lawyer agreed that although it was possible for them to do so, there was little hope of success.

Wiley appealed to the U.S. Public Health Service to see if it would investigate the disease. The health service replied, expressing a vague interest, sent a few pamphlets and cited a few references on radium, but offered no action. In addition, she pressured the state labor department to ask the public health service to intervene; however, nothing ever came of it.[21]

In desperation, Wiley approached two friends and colleagues from the National Consumers' League: Florence Kelley (Figure 25) and Dr. Alice Hamilton (Figure 26). Kelley was the general secretary and Hamilton the vice president of the national league. Both women were associates of Jane Addams (1860–1935), having worked with her at Chicago's Hull House, the famous pioneer community settlement house. And both were ardent social justice activists.

Florence Kelley (1859–1932), the daughter of a U.S. Congressman, had fought for decades for better working conditions for women, a shorter workday, a minimum wage, and the abolition of child labor. Her friends considered her an outspoken socialist, while her enemies believed she was a communist. Kelley was investigated several times by the FBI, and the agency's file on her noted in 1923 that she was "a radical all the sixty-four years of her life."[22]

Dr. Alice Hamilton (1869–1970), on the other hand, was a pioneer occupational physician. For decades, she combated the industrial hazards of lead, phosphorus, mercury, and a host of other toxic substances. She always was the champion of victims of industrial diseases, using a soft-spoken manner to make her points. Hamilton was the first woman appointed to the faculty of Harvard University as an assistant professor in the Industrial Hygiene Department of the university's School of Public Health. And her department chairman was Dr. Cecil K. Drinker.[23]

Figure 25. Florence Kelley.

When Wiley told Florence Kelley the fantastic, macabre story of the radium dial painters, Kelley was perplexed and horrified. She found it so shocking that she publicly denounced the situation as "cold-blooded murder in industry."[24] Alice Hamilton also was very concerned and interested in conducting an investigation of the situation.

To help Dr. Hamilton, Wiley visited John B. Andrews (1880–1943) of the American Association for Labor Legislation. As the secretary of the labor association, Andrews had worked for years with the U.S. Bureau of Labor Statistics to eliminate the use of white phosphorus in the nation's match industry. Now he was pursuing the end of its use in fireworks manufacturing.

When Andrews was told of the dial painters and how their mysterious disease was similar to phosphorus poisoning, he immediately wrote to Ethelbert Stewart (1857–1936),

Figure 26. Dr. Alice Hamilton.

the U.S. Commissioner of Labor Statistics, requesting that he invite Dr. Hamilton to conduct a special study of the situation. The Commissioner agreed; however, nothing was done for more than a year.[25]

In the meantime, Wiley continued to try to persuade others to study the dial painters in an effort to help them. In late 1924, she convinced the well-known statistician, Frederick L. Hoffman (1865–1946), to conduct an investigation. Hoffman, a long-time consulting statistician to the Prudential Insurance Company of Newark, was a former president of the American Statistical Association, and one of the founders of the American Cancer Society (Figure 27). He was particularly interested in occupational illnesses, especially the prevention of cancer.[26]

Wiley shared her findings with Hoffman, who was immediately struck by the number of young women who died and were sick, all of whom had little in common other

Figure 27. Frederick L. Hoffman.

than having worked at the same plant. The statistician felt the high level of morbidity and mortality among the young women could not be a mere coincidence.

To conduct his investigation, Hoffman developed a questionnaire and sent it to physicians and dentists who would have likely seen the dial painters. He also interviewed many of the women and the health professionals who had treated them.[27]

Hoffman quickly determined that the workers were indeed suffering from a new occupational disease. He was convinced that the cause of the disease was the chronic habit of the dial painters of pointing their paintbrushes with their lips, resulting in the swallowing of small quantities of radium, which were retained in the body.

Both Hoffman and Wiley wrote separately to the president of the U.S. Radium Corporation and confronted him with their findings. Arthur Roeder, however, continued to deny that radium was responsible for the women's illnesses. He invited each of them to personally inspect the radium plant. Hoffman accepted the offer and toured the facility. However, at the time, no dial painters were working there.

By January of 1925, Hoffman had completed his investigation, and he was invited to present his findings at the annual meeting of the American Medical Association, which was being held at the end of May. The paper then would automatically be published in the association's journal.

In early February, Dr. Hamilton again wrote to Katherine Wiley, indicating that she was still interested in conducting an investigation of the dial painters. Hamilton said she would write directly to the U.S. Bureau of Labor Statistics and ask them to commission her to conduct the study. However, a few days later she notified Wiley that the whole situation had changed. She indicated that Dr. Drinker would likely publish the Harvard team's report, which would make it unnecessary for her to conduct a study. Hamilton also wrote that if Roeder was "... stupid enough to refuse to let Dr. Drinker publish the report ...," she and Wiley would work behind the scenes to force the issue.[28]

Drinker had received an invitation to present the results of the team's study at the American Medical Association meeting at the same session where Hoffman was reading his paper. After obtaining a copy of Hoffman's paper, Drinker wrote to Roeder. He indicated that Hoffman's paper described the cases of jaw necrosis at the radium plant and that it specifically identified the company and its employees. He also wrote that the paper mentioned the Harvard investigation and noted that nothing ever came of it.

Drinker requested that Roeder immediately give him permission to publish the results of the study. He argued that it would clearly be in his best interest because it would show that he had "... done everything humanly possible to get to the bottom of the trouble ... ."[29]

Roeder wrote back and said he would study the matter and reply in a few days. However, after several weeks delay, he finally contacted Drinker and stated because of a lawsuit, which had been just filed against the corporation by a dial painter, he could take no action regarding the publication of the report, but had turned all of the material over to the company's lawyers.[30]

In the meantime, Alice Hamilton wrote to Katherine Drinker, Cecil's wife, indicating that Katherine Wiley had just informed her that Roeder had not only been telling everyone that the Harvard report exonerated him from all responsibility for the illnesses of the dial painters, but that he had even gone so far as to issue to the New Jersey labor department a forged report in their names. Hamilton asked Mrs. Drinker: "Do you not think that by misquoting you, using your name to bolster his cause, Mr. Roeder is failing in honorably dealing with you?"[31]

When Katherine Drinker showed the letter to her husband and Dr. Castle, they were indignant. If it were true that Roeder had behaved in such an outrageous manner, they agreed, they would be entirely justified in publishing their report without any further delay.[32]

To see whether Hamilton's account was true, Dr. Cecil Drinker immediately wrote to the New Jersey labor department, asking for a copy of the report they had been provided by the U.S. Radium Corporation. Upon receiving it, Drinker found that it had indeed been modified. Although the report given to the labor department was a literal transcription of some parts of the original report, it had been abstracted in such a way as to reverse the original report's actual conclusions.[33]

Drinker was so angry that he went to the radium corporation's headquarters in New York City to personally confront Roeder. Drinker told the president of the corporation to immediately provide a copy of the true report to John Roach of the New Jersey labor department or else he would publish the report. Roeder agreed.

A few days later, on May 30, Frederick Hoffman presented his paper at the American Medical Association's meeting. His paper, entitled "Radium (Mesothorium) Necrosis," concluded: "... we are dealing with an entirely new occupational affection demanding the utmost attention ... ."[34] The paper's findings were immediately picked up by the press.

Several weeks later, Dr. Drinker found that Roeder still had not given the original report to John Roach as promised. Instead Roeder's lawyer had forwarded it directly to the commissioner of labor. Frustrated, Drinker again wrote to the president of the radium company. He indicated that because Roeder had not kept his word, he was immediately arranging for the publication of the report.[35] Roeder and his lawyers threatened to sue for libel, but Drinker published it anyway.[36]

After more then a year's delay, the Harvard report, entitled "Necrosis of the Jaw in Workers Employed in Applying a Luminous Paint Containing Radium," finally appeared in print. Drinker, who was the founding and managing editor of the *Journal of Industrial Hygiene*, which was published by the Harvard Medical School, arranged for the report to be printed in the August 1925 issue of the journal.

While Hoffman was the first publicly to describe the new occupational disease, Drinker felt that he and his colleagues were the first to investigate it thoroughly. And when the Harvard report appeared in the journal, Drinker predated it as being submitted for publication on May 25, 1925, five days before Hoffman presented his paper.

Although Hoffman and the Harvard team's papers were important contributions, they did not provide conclusive proof that the cause of the new disease was radium. A more in-depth investigation of the dial painters would be needed.

The Medical Sherlock Holmes

By the end of 1925, a New Jersey physician and forensic pathologist would finally furnish the definitive evidence that the new occupational disease was caused by radium. The physician would spend the rest of his career following and studying the dial painters. He would write numerous papers about them and become the world's leading medical authority on the effects of radioactivity taken into the human body. His name was Dr. Harrison Stanford Martland (1883–1954) (Figure 28).

Figure 28. Dr. Martland (left) modeling a skull to help identify a homicide victim. Watching are Dr. Maurice Powers, Deputy Commissioner of the Royal Canadian Mounted Police, and, in the rear, Michael Frunzi, [Newark] City Hospital laboratory technician, February 1938.

Born in Newark, Martland received his medical degree from Columbia University in 1905. He completed his residency at Bellevue Hospital. Then he worked as an assistant pathologist for the Russell Sage Institute of New York City. In 1908, he was appointed pathologist at the Newark City Hospital, and a year later, he also became the pathologist for the City of Newark. In 1920, Martland was named the assistant county physician of Essex County, New Jersey. Five years later, he was appointed county physician (chief medical examiner), a position he would retain for nearly 30 years.[37]

During his long medical career, Dr. Martland performed over 15,000 postmortems. Among those he autopsied were the notorious gangsters "Dutch Schultz" and "Bugs" Moran.

As chief medical examiner, Martland was frequently involved in spectacular court cases. He was so successful at solving murders and in proving the innocence of accused persons that newspapers dubbed him the Sherlock Holmes of medicine.

The position of medical examiner would provide Martland with a unique opportunity to study the radium dial painters. He would have the full legal authority to investigate and autopsy all suspected victims of the new occupational disease who died in the county, and most of the dial painters who worked at the radium plant resided within its borders. And in his efforts to decide the exact cause of death, he would become a pioneer in the development of methods to determine the amount of radioactive material inside the body.

Martland first became aware of the mysterious disease experienced by the dial painters around 1923, when Dr. Craster, the director of the Newark health department, called his attention to the situation. When one of the painters died, Martland tried unsuccessfully to obtain permission to do an autopsy. Later, he saw several of the women at the dental offices of Drs. Davidson and Barry. However, he temporarily lost interest in the matter.[38]

By May of 1925, however, things dramatically changed. That month Dr. Martland examined two sisters, Margaret Carlough (1901–1925) and Sarah Carlough Maillefer (1889–1925), who were both critically ill. Both women started working as dial painters in 1917. And both had worked until they became ill. Margaret was employed until 1923, while Sarah had worked until March 1925. The women were now suffering from extensive jaw necrosis and severe anemia.

That same month, Martland saw Edwin D. Leman (1888–1925), the radium plant's chief chemist, who also was critically ill. Leman, who received his doctorate in chemistry from the University of Chicago, had processed radium for 14 years, spending the last four years working for the U.S. Radium Corporation. Although he never painted dials, he handled large concentrations of radium. As a result, the thumbs and fingers on both of his hands were seriously scarred.

A year earlier, when the Harvard public health team examined Leman, he scoffed at the possibility of any future health problems.[39] However, he soon began experiencing frequent fainting spells. In May, he collapsed and was hospitalized. Despite several blood transfusions, his condition continued to worsen. When in early June, he died, at the age of 36, the diagnosis was severe anemia.

To determine the exact cause of death, Martland conducted an autopsy. With the help of a physicist from the U.S. Radium Corporation, he determined the amount of radium in the chemist's organs. His lungs were found to be particularly radioactive, indicating the inhalation of radioactive dust. In addition, radium was found in his spleen and bones. The chemist's death clearly was the result of the radium taken into his body, as well as his exposure to external radiation given off by the element.[40]

While they were still alive, Dr. Martland also measured the amount of radium in the bodies of Carlough and Maillefer. And when Mrs. Maillefer died, he conducted an autopsy on her. He again found high concentrations of radium in the woman's spleen and bones, as well as in her liver. Martland attributed her death to anemia caused by her occupational exposure to radioactive substances.

To further verify his findings, Martland measured the amount of radium in several other living dial painters. Although the women appeared to be in good health, they all were found to have radium in their bodies—the radioactive gas radon, which is released as radium decays, was found in their exhaled breaths; likewise, gamma rays emitted by the element and its decay products also could be detected.

Martland quickly wrote up the results of his findings and submitted them for publication to the American Medical Association. Although his paper was accepted, it would not be published for many months. To get his findings into the medical literature as quickly as possible, he presented a summary of his results at the October meeting of the New York Pathological Society. Papers read at the meeting were automatically published in the proceedings of the society.[41]

On December 5, 1925, Martland's paper, entitled "Some Unrecognized Dangers in the Use and Handling of Radioactive Substances," appeared in the *Journal of the American Medical Association*. In it, he wrote:

> This report is published now as a warning that when long lived radioactive substances are introduced into the body ..., death may follow a long time after, from the effects of constant irradiation on the blood-forming centers. ... The cases reported in this study demonstrate the importance of, and the possible tragedy due to, the late and largely underestimated effect of radioactivity.[42]

Although the article would eventually become a medical classic, its findings initially were attacked, ridiculed, or ignored. Even Dr. Morris Fishbein (1889–1976), the well-known and powerful editor of the *Journal of the American Medical Association*, was skeptical. Radium was just too widely used, and too miraculous, to be considered deadly.[43]

One scientist who refused to recognize the new disease was Frederick B. Flinn (1876–1957) of the Institute of Public Health of Columbia University. Flinn started his career as a metallurgist and was director of several mining companies for a number of years (Figure 29). Later, he became interested in physiology and worked for the U.S. Public Health Service. In 1922, he was hired as an instructor at Columbia. After completing his doctorate degree in 1923, he was promoted to assistant professor of physiology, specializing in industrial hygiene.[44]

Flinn also was a consultant for the U.S. Radium Corporation. The corporation had hired him in March of 1925, hoping he would refute the Harvard public health team's report. To study the effects of radium, Flinn conducted a number of animal experiments. And with the help of the firm, he examined practically every radium dial painter working in America and had other dial painters examined for him in Europe.

Figure 29. *Frederick B. Flinn.*

In December of 1926, Flinn published the results of his work in the *Journal of the American Medical Association*. The article, entitled "Radioactive Material an Industrial Hazard?," failed, however, to cite the earlier paper by Hoffman, and the definitive work of Martland, despite the fact that both were published in the pages of the same journal. Flinn only cited one article dealing with the new disease—that of the Harvard public health team.

Flinn's article, which Hoffman called "more bias than science,"[45] brashly concluded:

> ... an industrial hazard does not exist in the painting of luminous dials. The only evidence contrary to this conclusion rests on the fact that five employees at the Orange plant of the United States Radium Corporation have died from some cause that cannot be determined at this date.[46]

In reaching this judgment, the industrial hygiene expert argued that if the deaths were caused by radium, other cases should have appeared at other dial-painting plants, because employees used the same materials, and they also pointed their brushes in their mouths.

However, in June 1926, a month after Flinn submitted his paper for publication, a dentist referred a woman to him for treatment of a condition suspected to be the result of industrial exposure. The woman was experiencing terrible teeth and jaw problems. In addition, she suffered from vague pains in her legs, and several years earlier, while at a dance, she had tripped and fractured a leg without actually falling to the floor.

Flinn carefully questioned the woman about her former employment, and found that she had previously worked as a dial painter for the Waterbury Clock Company of Waterbury, Connecticut. The company's dial-painting department had been started with the help of the U.S. Radium Corporation, which had trained its employees and also had supplied the firm with its radium paint.

Flinn found that eight years earlier, the woman had painted dials for only 14 months. And when he tested her for radium, he found the telltale sign of radon in her breath. From the amount of radioactive gas exhaled, he estimated her body contained a large quantity of radium.

Having seen the new case, Flinn was in a quandary as to what to do about his paper, which was already scheduled for publication in the prestigious medical journal. He asked the head of his department for advice and was told not to withdraw the paper as his "work up until then was alright [sic]," and until he "was positive that the case sent in was a radium case" he "could do nothing but later on could continue his findings."[47]

Several months later, Flinn saw a second dial painter from the same Connecticut plant who also was clearly suffering from radium poisoning. Despite the overwhelming evidence of the new disease, Flinn still refused to withdraw the paper. Later, he would regret his decision.

In May 1927, Flinn published a second paper, "A Case of Antral Sinusitis Complicated by Radium Poisoning," describing the medical condition of the first Connecticut dial painter he had seen and briefly mentioning the second case. This time, he meekly wrote: "In view of the evidence in front of me, I feel that radium is partially if not the primary cause of the pathologic condition described."[48]

In a third paper, entitled "Some of the Newer Industrial Hazards," published in January 1928, Flinn discussed radium poisoning. He reported how he had changed his mind about the new occupational disease after seeing the Connecticut dial painters. He wrote: "These two cases, appearing thus in another state from the first cases, have caused me to suspect that radioactive material is at the bottom of the trouble even if the mechanism by which it is caused is not altogether clear and not previously suspected."[49]

In February 1929, Dr. Martland would take Flinn to task. In a special review article on the new occupational disease appearing in the *Journal of the American Medical Association*, Martland presented a detailed description and history of radium poisoning. He reviewed Flinn's three articles and his changing views.

In discussing Flinn's third article, Martland took particular issue with his statement that the radium poisoning cases from other dial-painting plants "were not previously suspected" and that the cause of the disease was "not altogether clear." Martland wrote:

> It is worthy of note that not only were these cases suspected by Hoffman, [and] Castle and the Drinkers in 1925, two and one-half years before these statements by Flinn, but the disease was accurately described in all its features by me and my associates two years before Flinn finally made the foregoing admissions.[50]

While Flinn had ruined his own credibility (Florence Kelley, for example, called him the "fraud of frauds"),[51] Martland increasingly gained national and international fame for his work.

In October 1931, Martland published another paper on the radium dial painters, which again would become a medical classic. The paper, entitled "The Occurrence of Malignancy in Radioactive Persons," appeared in the *American Journal of Cancer*. In the large, comprehensive article, which ran 82 pages in length, Martland reviewed the early victims of radium poisoning, who tended to die from severe anemia and extensive jaw necrosis. He then presented in detail the newer cases, in which former dial painters suffered from osteogenic sarcomas (malignant bone tumors) and other rare cancers.

Based on his review, Martland concluded:

> I am now of the opinion that the normal radioactivity of the human body should not be increased, strongly presuming that increased amounts of radioactivity may produce, over a number of years, malignancy. ... From my experience, it is impossible to state what is the greatest amount of radioactivity the body can safely carry. ... Theoretically the exposure to, or the use of any radioactive substance that will increase the normal radioactivity of the body is dangerous.[52]

For decades, Dr. Martland continued to follow and study the dial painters. In 1925, with the help of one of the painters, he compiled a list of 63 workers who were known to have been employed at the radium plant for a long time. Over the years, when he heard of a woman suffering from a mysterious illness or death, he would consult his "list of the doomed." All too frequently, Martland found her name there. With terrible, almost mathematical regularity, the women developed symptoms and died from radium poisoning.[53]

In December 1951, *Life* magazine published an article on the death of the 41st New Jersey dial painter and the work of Dr. Martland. The article prominently displayed a photograph of the famous medical examiner, sitting at his desk at the Newark City Hospital, checking his "list of the doomed," as he had done for many years (Figure 30).[54]

During Martland's career, he was accorded numerous awards, honors, and positions. He was elected to many medical societies and was on the editorial boards of several medical journals. For a number of years, he served as a member of the Nobel Prize nominating committee in medicine. He also was offered many posts at medical centers,

Figure 30. Dr. Martland at his desk with his "List of the Doomed," Life *magazine, December 1951.*

universities, and municipalities. Mayor Fiorello La Guardia (1882–1947), for example, offered him the position of medical examiner of New York City. Martland, however, chose to stay in Newark.[55]

Shortly before his death in 1954, Martland received perhaps his greatest honor. The new municipal hospital erected to replace the old Newark City Hospital, where he had performed his groundbreaking work, was named the Harrison Stanford Martland Medical Center.[56]

In sharp contrast, when Frederick Flinn died several years later in 1957, he specifically instructed his heirs to destroy all his notes and records, presumably so no trace of his involvement with the U.S. Radium Corporation could ever be investigated.[57]

Now that the new occupational disease was finally recognized, the radium dial workers, and their families, would be faced with the very difficult task of obtaining compensation for their terrible injuries.

Chapter 6

In Search of Justice

There must be some power in New Jersey, either in the courts or in public opinion, strong enough to bring the U.S. Radium Corporation before a tribunal promptly and to compel it to answer the claims of these women.[1]

Walter Lippmann, 1928

The First Lawsuit

Beginning in 1925, and continuing for more than a decade, the U.S. Radium Corporation would be inundated by lawsuits filed by the dial painters and their families. At first, the company would quietly settle the suits out-of-court to avoid the adverse publicity. Later, as the number of suits increased, the company would bitterly fight in court and attempt to delay and drag out the legal proceedings for years. Eventually, a New Jersey federal court would rule in the corporation's favor and bar the women from seeking any further damages from the company.

The first dial painter to file a lawsuit was Margaret Carlough. Margaret, like her dead sister, Sarah Maillefer, also was suffering from radium poisoning. In March of 1925, she sued the U.S. Radium Corporation for $75,000, charging that she had been poisoned and totally incapacitated from her employment.[2]

Several others also joined her lawsuit: the family of Hazel Kuser, which was trying to recover the costs of her medical and funeral expenses; and Dr. Joseph Knef, who was seeking payment for dental care he had provided to Margaret and her sister.

In May of 1926, shortly before the case was to go to trial, the U.S. Radium Corporation settled the suit out-of-court. Without assuming any legal responsibility, the company agreed to pay a total of $13,000. But the settlement was little more than a hollow victory. Margaret had died, and much of the money went to her lawyer, who charged a contingency fee of 45% for taking the case.[3] The same attorney represented all plaintiffs in the suit.

Margaret's family received $9,000, with $4,050 of it going to the lawyer. The Kuser family, which spent almost $9,000 in expenses, received a token $1,000, with $450 going to the lawyer. Dr. Knef received $1,000, again with $450 going to the lawyer. In addition, the lawyer received $2,000 of the settlement for legal expenses and court costs.[4]

After receiving his settlement, Dr. Knef contacted the U.S. Radium Corporation and requested a meeting with the company's board of directors. Knef said that he had an important proposition for them. And a conference was scheduled at the corporation's New York City headquarters.

At the meeting, which was attended by the corporation's directors, as well as the president, treasurer, and legal counsel of the company, Dr. Knef stated that the dental care he had provided to several dial painters was a severe financial drain on his practice; and the dentist said that it was necessary for him to receive compensation for it. Knef then indicated that he would get paid by "either one side or the other," and he wanted to know whether the company was going to "play ball" with him or not.[5]

Knef indicated that he could make himself of great value to the corporation in several ways, including influencing the dial painters not to sue the company and testifying for the company if any suits were brought against it. The dentist also said that he was an expert in treating radium necrosis and that he had devised ways of stopping the spread of the disease. He claimed that he had designed a secret treatment involving a machine that could extract some of the radium from the bones of living patients. Knef said that the longer that he kept the women alive, the more likely they would get other employment, move out of the vicinity, and not be heard of any further, or die from other causes.[6]

When the company's chairman of the board questioned Knef about the specifics of his proposal, the dentist said that the corporation should pay him the sum of $10,000 for care he had already provided. In addition, Knef stated that he wanted the company to pay him a set fee for all future dental care he would provide to the dial painters. Last, Knef asked the company to furnish him with a list of all of its former employees so that he could contact them for examination and treatment.

The dentist threatened that if the company did not accept his proposition, he would sue some of the dial painters to collect his fees. Knef declared that he could testify that a particular case was the result of either radium poisoning or periodontal disease and that "he could get away with anything he wanted as these things were little understood."[7]

Knef was asked whether his proposition meant that if the company paid him, he could save it a great deal of expense and trouble, and that if it refused, he was in a position to make a lot of trouble for it and would do so. The dentist replied that yes he could help and that his proposition was "not blackmail either."[8]

At the end of the meeting, the corporation's president, Arthur Roeder, called the dentist's proposition absolutely immoral and said the company would have nothing whatsoever to do with it. Knef replied: "Immoral is it? Is that final?" Then he left.[9]

Following this encounter, Roeder and all of the others attending the meeting wrote out sworn statements describing Knef's proposal. The company then sent the affidavits to the New Jersey state dental board, hoping that it would revoke Knef's license. After reviewing the statements, the state board said its charter would not permit it to take any action. But it referred the matter to the ethics committee of the New Jersey dental society. The ethics committee refused to take any action. Instead, it referred the matter to the society's Newark chapter, of which Dr. Knef was a member. The chapter also took no action, and eventually the matter was dropped.[10]

Despite Roeder's strong rebuff and possible sanctions by the dental society, Knef continued to telephone and write various officers of the radium company asking for money. The company, which kept detailed records of all of his communications, continued to refuse all of his demands. For a time, the firm even went so far as to hire a private detective to investigate Knef's activities. Eventually, the dentist ceased his efforts.

Knef continued to practice dentistry, and over the years, repeatedly claimed that he was the first to discover and investigate radium poisoning.[11] He also falsely told various newspapers that he had conducted many studies of the dial painters. After reading one article that quoted Knef, Frederick Hoffman called him "a self-advertising dentist, who desired publicity." Hoffman also said that Knef was "not a scientific man and was not competent" and that he had "not made the slightest scientific investigation" of the disease.[12]

At the end of June 1926, Arthur Roeder was forced out of the U.S. Radium Corporation by its board of directors because of his ineffectual handling of the dial painter crisis. After leaving the company, Roeder quickly became the executive vice president of the American Linseed Oil Company of New York. Later, he would become the president and chairman of the board of the Rockefeller-controlled Colorado Fuel and Iron Company, that state's oldest and largest industrial firm. He would remain with that company for many years until he retired.[13]

The Case of the Five Women Doomed to Die

If the directors of the U.S Radium Corporation thought they had disposed of the crisis, they were mistaken. The company would soon face an even greater challenge. In 1927, five former dial painters would file a lawsuit against the firm for a total of $1,250,000 in damages. The resulting trial would arouse worldwide interest and sympathy for the women. Newspapers would call it "The Case of the Five Women Doomed to Die."

The women, Grace Fryer, Quinta McDonald, Albina Larice, Edna Hussman, and Katherine Schaub, were hired by the radium corporation in 1917. They had painted dials for one or more years. And when they left the company, all of the women were healthy. Several years later, however, they began experiencing rheumatic-like pains in their lower extremities including their feet, knees, legs, and hips. Several women also developed tooth and jaw problems. Eventually, all would develop and die from crippling bone sarcomas (cancers).

Figure 31. *Grace Fryer.*

Grace Fryer (1899–1933) (Figure 31) started working for the radium company when she was 18. She was employed for three years and ten months. Several years after leaving the firm, she began to suffer pain in one foot. Soon the pain intensified and spread to her back. She saw several physicians who treated her for rheumatism and tuberculosis of the spine. Despite the treatments, her condition worsened. Radium was slowly eating away her spine, and her vertebrae were collapsing and telescoping onto each other. As a result, she could not walk without the help of steel braces on her back and feet. In time, she developed problems with her arms; she could not raise the right nor bend the left. Grace also suffered from necrosis of the jaw that required 19 dental operations. In 1925, she was seen by Dr. Martland, who diagnosed her as suffering from radium poisoning.[14]

Mrs. Quinta McDonald and Mrs. Albina Larice were sisters. They had worked as dial painters along with another of their sisters, Amelia "Mollie" Maggia. Mollie, however, had mysteriously died in 1922, although the doctors listed the cause of death as syphilis.

Quinta Maggia McDonald (1900–1929) began working for the radium company at 17. Of the five women, she was employed for the shortest time, 15 months. But when she was at the company, she was said to have frequently eaten the paint, commenting many times on its gritty taste. After getting married, she left the firm. Shortly after giving birth in 1923, she began having health problems. At first, she suffered severe pain in one hip. Later, the pain spread to her knees, left arm, and right wrist. Soon she began limping. In 1925, she also was diagnosed by Dr. Martland as suffering from radium poisoning.[15]

Albina Maggia Larice (1894–1946) was the oldest of the five women; she was hired at the age of 23. She worked for the company for 20 months. In late 1925, she felt pain in her knee, which then spread to her hip. Eventually, her hips were locked in place and she could not move her legs more than a few inches. One of her legs also shrank, and it became very difficult and painful for her to walk. In 1925, she was diagnosed by her physician as suffering from radium poisoning.[16]

Mrs. Edna Hussman (1901–1939) worked the longest time for the company. She was hired at the age of 16, and worked for more than five years. After leaving the firm, she also worked as a dial painter for another radium company for about a year. As did Albina, she had pain in one knee. Later, the pain spread to a hip, her other leg, and her arms. She went to a hospital and was found to have suffered a spontaneous fracture of the femur (thighbone). Her leg and hips were put into a plaster cast for a year. Although her leg appeared to have healed, she soon noticed that it was becoming shorter. In time, her left leg became three inches shorter than the other, as the radium was slowly destroying her thigh bone. Although she was suspected of suffering from radium poisoning in 1925, the diagnosis was not confirmed until 1927 by her surgeon.[17]

Katherine Schaub (1902–1933), the youngest of the five women, started work at the age of 15. She worked for the radium company for almost five years, first as a dial painter and later as an instructor. After leaving the firm, she also worked for another radium company for 8 months. Katherine was said to have infrequently pointed the brush in her mouth. However, in 1923, she began to have problems with her teeth. After two teeth were extracted, the sockets remained inflamed and small pieces of bone came out. She was treated for jaw necrosis by several dentists. Later, she began to experience neuralgia-like pains throughout her body. Eventually she developed severe pain in her left knee, and she began limping. In 1925, Dr. Martland found she was suffering from radium poisoning (Figure 32).[18]

After Grace Fryer was diagnosed with radium poisoning, she asked her attorney to take her case against the U.S. Radium Corporation. He refused, explaining that it would take considerable money and that there was little likelihood of success. Grace then tried to find other lawyers who might be interested in the case, but all declined.

Figure 32. The five doomed women. Left to right: Quinta McDonald, Edna Hussman, Albina Larice, Katherine Schaub, and Grace Fryer.

In the spring of 1927, she approached a small law firm in Newark to see if they would help her. Although it appeared to be a lost cause, the firm's junior partner, a young Yale-trained lawyer, Raymond H. Berry (1897–1971), decided to accept her case.[19] Berry knew that legal proceedings against the large radium corporation would be very difficult, especially since there were so few legal precedents to point the way.

Berry started gathering evidence by reviewing the files of the New Jersey Consumers' League and the research work of Frederick Hoffman on the radium victims. He also began interviewing many of the former dial painters, their physicians, and dentists.

On May 18, 1927, Berry filed Grace Fryer's lawsuit in the New Jersey Supreme Court. A month later, Katherine Schaub joined the suit. She was soon joined by the three other women.[20]

In their suit the women charged that the U.S. Radium Corporation had poisoned them, and that despite its knowledge that the radioactive element was harmful, the company had not warned them or provided them with adequate safety measures. The suit also alleged that the corporation had suppressed the Harvard University's public health team's report, that it had hired Frederick Flinn to refute the team's work, and that the company had intentionally misled the New Jersey labor department concerning the report's findings.[21]

The radium corporation's lawyers said the company was not negligent and that the women were not entitled to sue. They claimed that the New Jersey statute of limitations, which specifies that a suit for damages must be brought within two years after the inception of a disease, had run out.

Berry countered, arguing that the statute applied from the moment the women had learned the true cause of their disease, and not from the time they developed symptoms, which in the case of this new occupational disease occurred years after the women had left the company. Berry also charged that the radium company had purposely misrepresented scientific opinion and organized a campaign of misinformation that prevented the women from suing.

To stop the company from using the statute of limitations defense, Berry transferred the case to chancery court in July 1927. That court would decide whether the statute applied or not. If the court ruled the statute defense was valid, the case would end. If, however, it ruled the statute did not apply, the women's case would then proceed in the supreme court. The first hearing was scheduled for the beginning of 1928.[22]

Before the hearing, to obtain key corroborating evidence for the case, Berry had the body of Amelia Maggia exhumed. Her remains would be used to determine whether the mysterious illness she died from was caused by radium.

Five years after her death, on October 15, 1927, in the presence of her father, numerous physicians and scientists representing the women and the U.S. Radium Corporation, as well as members of the press, Mollie Maggia was exhumed. Her body was taken to a local funeral home where it was autopsied, and samples of bones and tissues were taken for analysis.[23]

The autopsy quickly revealed that Mollie had not died from syphilis. When the tests came back, every piece of bone, as well as every tissue, was found to be highly radioactive. Based on the amount of radium in the samples, it was estimated that her body contained a total of 48.4 micrograms of the radioactive element (an amount approximately 500 times greater than the safety standard that would later be developed). Mollie had unquestionably died from radium poisoning.[24] And the disease was now slowly killing her sisters.

The women's first court hearing took place on January 12, 1928, in Newark. The one-day session would prove to be emotionally wrenching. When the five women entered the courtroom, their appearance was shocking. The young women were destroyed: They were disfigured by jaw necrosis, and crippled by rotting spines and shrunken legs. Several could not walk and had to be carried to the witness stand. One could not even raise her arm to take the oath.

At the hearing, Mrs. Hussman, Miss Fryer, and Mrs. McDonald were called to testify. The women told how they had been taught to point the paintbrush in their mouths in order to obtain a fine point. They also described the severe health problems that they had developed. As they testified, friends and spectators in the courtroom openly wept. At the end of the day, the case was adjourned until April.[25]

The second hearing took place on April 25–27. On the first day, Katherine Schaub and Mrs. Larice testified. They were followed by several physicians and a toxicologist who had conducted the autopsy of Amelia Maggia. They reported their findings that indicated the young woman had clearly died from radium poisoning. The physicians also discussed the prognosis of the five women. All agreed they were suffering from fatal radium poisoning.[26]

During the second day, among those to testify were Dr. Harrison Martland and Arthur Roeder. Martland was asked about how he had gotten involved with the radium cases. He also was asked to describe his research findings concerning radium poisoning. When Roeder testified, he was questioned as to whether he was aware that dial painters pointed their brushes orally. The former president of the company said that although he had frequently visited the radium application plant where the dial painters worked, he had never noticed the practice.[27]

On the third day of the session, Dr. Sabin von Sochocky, Frederick Flinn, and Katherine Wiley were called as witnesses. Dr. von Sochocky was asked about the early history of the U.S. Radium Corporation. He also was asked whether he had seen Grace Fryer put the paintbrush in her mouth and whether he had instructed her not to do so. The doctor explained that he had indeed told her not to do so, but that his reason for this admonition was his concern about sanitation and not about the radium in the paint, which he considered to be infinitesimal.

When Frederick Flinn testified, he was asked about his involvement with the U.S Radium Corporation. Flinn described his research work and how he had slowly changed his mind concerning the existence of the new disease over time.

The last witness was Katherine Wiley from the New Jersey Consumers' League. She described how the league had become involved with the dial painters at the request of the Orange, New Jersey, public health department. She also discussed the league's efforts to investigate the cause of the mysterious disease.[28]

After Wiley's testimony, Berry rested the plaintiffs' case. The young attorney thought that the defense would immediately begin presenting its case. The radium corporation's lawyers, however, asked for an adjournment. They indicated that they wanted to call several expert witnesses. They claimed that, unfortunately, some of the witnesses would be in Europe, while others were professors who would not be available to testify until the fall. The judge, after reviewing his court calendar, also said that he would not have time to hear the case until the end of September.[29]

Berry strongly objected. He believed that the U.S. Radium Corporation was stalling and feared that some of the dial painters might not live until then. The judge, however, quickly overruled his objection.

By this time, the news media were increasingly interested in the plight of the women. Newspaper headlines screamed: "Radium Case Off Till Fall," "Five Poisoned Women Face Court Delay," and "Suit of Five Women Facing Death Adjourned to Seek Evidence."[30]

Numerous articles and editorials appeared criticizing the U.S. Radium Corporation and the court. The most influential editorial, however, was published on May 10, 1927, in the New York *World*, at the time one of the nation's largest and most powerful newspapers. It was written by Walter Lippmann (1889–1974), a crusading journalist who would become the nation's most widely respected political columnist.[31] Lippmann had been informed of the women's lawsuit almost a year earlier by Dr. Alice Hamilton.

In the editorial (Figure 33) entitled, "Five Women Doomed to Die," Lippmann reviewed the history of the case and scathingly wrote:

> ... this is one of the most damnable travesties on justice that has ever come to our attention. It is an outrage that the company should attempt to keep these women from suing. It is an even greater outrage that Jersey justice should tease them along for fourteen months before deciding whether they have a right to sue.[32]

The next day, the radium corporation's consultant and star witness, Frederick Flinn, told reporters that he had tested the five women and had found no trace of radium in their bodies. Later, the industrial hygienist told the press that with proper care, all of the women might become better and regain a measure of health that would give them a fairly long life.[33]

Lippmann objected to Flinn's statements. In another editorial, he wrote that Flinn's remarks had "all the appearance of being timed to support the argument of the lawyers for another four month's delay" and that they do not "make the case one bit less horrible."[34]

Frederick Hoffman also strongly disagreed with Flinn's statements. The statistician told the press that in his opinion, none of the women would live much longer than five years and that most of those years would be spent in great agony.[35]

Even Marie Curie was drawn into the debate. In a rare interview, the eminent scientist addressed the women's situation. Although she offered her sympathies, she provided no cure, and no hope. She sadly explained: "I see no hope for them. While I am not a physician, my experiments with radium convince me that if the poison is taken internally, it is practically impossible to destroy it. If the radium had entered the body through the lungs, there is a slight hope—but not now."[36]

Marie Curie did, however, take the opportunity to criticize the U. S. Radium Corporation. She said that she was surprised that a factory with such bad hygienic conditions would exist in the United States. She declared: "I have never heard of a case of this kind in France—not even in wartime when countless factories were employed in work dealing with radium—nor did I ever hear of a similar case in Switzerland, where there are so many watch factories."[37]

The women's case also was becoming a political issue. The Socialist Party Presidential Candidate, Norman Thomas (1884–1968), who was often called the "conscience of America," referred to the dial painters in a campaign speech he gave in Newark. Tho-

Figure 33. An editorial cartoon from the New York World *criticizing the legal delays endured by the five doomed women, May 14, 1928.*

mas called the women's plight a "vivid example of the ways of an unutterably selfish capitalist system which cares nothing about the lives of its workers, but seeks only to guard its profits." He went on to say that these women on the "threshold of a horrible death, seek small measure of redress from the courts. Immediately." However, he declared: "every legal device is employed by the corporation and insurance company's lawyers to cheat them even of compensation."[38]

By this time, because of the enormous criticism, the chancery court judge hearing the women's case told their attorney, Raymond Berry, not to wait for his ruling, but rather to proceed with the case in the state supreme court. Berry agreed, and a court date was set for early June.[39]

Before the trial began, a young, public-spirited federal judge, who had no official connection with the case, volunteered his services as a mediator. Judge William Clark (1891–1957) of the U.S. District Court of New Jersey, had become interested in the women's plight after reading Lippmann's May 10th editorial.[40] Judge Clark found that he knew several of the people involved in the case, and he thought he might quickly obtain a settlement. The judge was concerned that even if the women's lawsuit were successful, the company would likely appeal, and the case might drag on for years without the victims receiving any relief.[41]

The judge's efforts were successful, and on June 4, 1928, the case was finally settled out of court. Without accepting any legal responsibility or liability, and supposedly because of humanitarian concerns, the U.S. Radium Corporation agreed to pay each of the women a lump sum of $10,000, and provide an annual pension of $600. The company also agreed to pay all past and future medical expenses connected with their injuries, providing they consented to periodic examinations by three physicians, who were to certify whether or not they were rid of radioactivity. Finally, the company agreed to pay all of the women's legal costs and expenses.[42]

This unique settlement would be the first publicly known instance in which persons injured by radioactivity received compensation. Reactions to the settlement varied. The women were generally satisfied, although at least one felt the amount of money was too low.[43] Raymond Berry expressed his opinion, saying: "While the action of the company in settling may be less humanitarian than that of a jury …, it is quicker and therefore more satisfactory in view of the possibility of these women dying at any time."[44]

The U.S. Radium Corporation's new president, Clarence B. Lee, claimed that although the company had a perfect legal defense, it had settled because it could not obtain a fair trial due to a cleverly designed publicity campaign by the consumers' league and the press. He denied the women were radioactive, and he said the condition they suffered from was the result of some other cause. He also bitterly blamed the victims. In a letter to the health commissioner of New York City he wrote:

> The [dial painting] work was easy, the operators well paid, and as conditions have turned out, we unfortunately gave work to a great many people who were physically unfit to procure employment in other lines of industry. Cripples and persons similarly incapacitated were engaged. What was then considered an act of kindness on our part has since been turned against us, as all previous employees, regardless of what they may have been suffering from or are suffering from at the present time, in the minds of the general public can be attributed to 'radium poisoning.'[45]

Newspapers and magazines hailed the settlement. The media felt they had helped the women receive a measure of justice before they would die from the terrible new disease. *Time* magazine, for example, declared: "Newspapers took these five dying women to their ample bosoms. Heartbreaking were the tales of their torture. Publicity hastened the case to trial through the lagging courts."[46]

The general public also was glad to see the case settled. People from all over the nation and the world sent the women letters of support and good wishes. The women also received numerous pieces of literature from faith healers and quacks promising to cure them.

The case also had other important repercussions. For the first time, the news media began to raise serious doubts and even fears concerning the use of radium and the developing new science of radioactivity. An editorial, entitled "Whose Responsibility?," in *Scientific American*, asked:

> ... who is to blame for the horrible blunder? Surely the girls were not. Are the employers? Years ago they should have heeded the warnings of general scientific information. Evidently they took these warnings with a large grain of salt ... . Perhaps it is a case where the greatest share of responsibility is no one's. A new field—radioactivity—was being explored; still is, in fact. Man blundered into a pitfall. He creates many engines which he has not learned fully to control.[47]

The Orange, New Jersey, *Daily Courier* took an even darker view. In an editorial, entitled "The New Death," it wrote:

> Edgar Allen Poe in all his weird stories never utilized a theme more harrowing than that of death by radium.
>
> Here we have the most mysterious of all elements; hailed by the scientific world as a new force of such magnitude as staggers the imagination turning like a boomerang to slay in a terrible way those innocently sought to employ it in the industrial arts.
>
> Experience teaches men how to control and utilize the discoveries which they coax from Nature but often it is at great cost. It was so with the first use of the X-ray. Many a man was frightfully burned and fell martyr to the cause of science until these uncanny rays were understood.
>
> We do not understand much about radium as yet. Apparently it is energy concentrated as we have never known energy to be concentrated. Apparently the tiniest particle of it has a power so great that if suddenly released it would move mountains. We seem to have stumbled across something more potent for good or evil than all other discoveries put together.
>
> It is all too amazing, too awful to understand and one cannot help but view with dread the very thoughts of the dangers that may be in store for the world as a result of the possible unleashing of this power.
>
> Many an imaginative person has feared that the day would come when man would discover some secret of the universe, the knowledge of which might place in his hands the power to wreck the world. As one reads of the slow death that is gradually drawing these young women to itself,

one wonders what trend the further investigation of this element may take and what disclosures may be in store. May we not believe that a power so great may yet be made subject to control? Or has man met his master at last?[48]

Building on the publicity given to the settlement, in July of 1928, Dr. Alice Hamilton, Florence Kelley, and Walter Lippmann appealed to the Surgeon General of the U.S. Public Health Service, Dr. Hugh S. Cumming (1869–1948), to sponsor a conference on the use of radium in the nation's industry. Originally, when the federal public health service was asked to investigate radium poisoning, it declined. It considered the new disease an occupational problem and therefore subject to the U.S. Department of Labor. The department's Bureau of Labor Statistics had started an investigation in 1925, but quickly discontinued it, because it needed greater technical resources. It would not begin a major study of the disease until the fall of 1928.[49]

Hamilton, who had previously been a member of a national public health service conference on the introduction of tetraethyl lead gasoline, believed that a similar conference could be helpful in investigating the use of radium in industry. Working with numerous other health care experts, she sent a letter to the surgeon general requesting a conference on the topic. The letter was signed by her and 23 other prominent individuals, mostly physicians.[50] After receiving the letter, the surgeon general quickly agreed.

The one-day conference was held in Washington, D.C., at the end of December 1928. It was attended by the surgeon general, the commissioner of labor statistics, Dr. Hamilton, Mrs. Kelley, Dr. Martland, and Frederick Flinn, as well as numerous representatives from the nation's watch and radium industries.

At the conference, the members agreed that most of the deaths from radium were caused by the practice of brush pointing, but others contended that any use of radium was hazardous. Ethelbert Stewart, the commissioner of labor statistics, called the use of radium on clock and watch dials "purely a fad." And he asked: "Do you want to go ahead with the use of a thing which is so useless, which has so little utility, which has, in spite of everything you can do, an element of serious danger in it?"[51] However, Richard B. Moore (1871–1931), former chief chemist of the U.S. Bureau of Mines, argued that all industries had health hazards and that if industries were shut down because of them, "we would shut down every industry."[52]

At the conclusion of the meeting, it was agreed that the U.S. Public Health Service should form two committees: one to conduct a field survey of the industry, and a second to codify the best known methods of protection.

Eventually, the public health service conducted studies of the radium industry, and it published several papers on the topic; little else came of its efforts. Dr. Martland summarized his feeling about the conference and the studies, writing: "The great trouble with most investigations is that they always start after the harm has been done."[53]

In June 1929, the "five doomed women" would again be covered intensely by the press. On the first anniversary of their settlement, newspapers and magazines noted with surprise that all of the women were still alive. They speculated that perhaps the radium within them was no longer radioactive.[54] A few of the newspapers, however, turned against the women because they had outlived the prophecy of their imminent deaths. The papers implied that perhaps they had fraudulently schemed to get money from the company.[55]

In time, however, the women would indeed suffer horribly and die from the effects of the radium. In 1929, a year and a half after the settlement, Quinta McDonald died at the age of 34.[56] She was followed in 1933 by Katherine Schaub, 30, and Grace Fryer, 35.[57] In 1939, Edna Hussman, 37, died.[58] Last, in 1946, more than 18 years after the settlement, Albina Larice died at the age of 51.[59]

Justice Denied

After the settlement, other former dial painters filed lawsuits against the U.S. Radium Corporation, and several were successfully settled out-of-court. Eventually, however, in the case of *La Porte v. U.S. Radium Corporation*, the federal court would bar the women from seeking any further damages from the company.[60]

Irene Corby (1900–1931) started working as a dial painter for the U.S. Radium Corporation when she was 17 years old. She worked for the company from May 1917 until December 1918. She also worked for the firm for a few weeks in 1920. When she left the company she was healthy, and she quickly forgot about the job.

In 1921, Irene married Vincent La Porte. After the marriage, Mrs. La Porte remained in good health until the fall of 1927. At that time, because of the publicity given to the five doomed women, she began to fear that she might be suffering from radium poisoning. Like the other women, she was beginning to experience trouble with her teeth and jaws.

At first, she was so fearful of having the new disease that she refused to see a dentist about the problem. However, after several of her teeth fell out, she decided to seek care. In the spring of 1928, she saw a dentist who treated her for jaw necrosis. After several months, her jaw appeared to heal.

Irene soon began to have continuous pain in her face and jaw. She went to a physician and told him that she believed she was suffering from radium poisoning. Around 1930, she also began to have pains in her legs and joints. Later that year, she developed a swelling in her abdomen. And in October, she was seen by Dr. Harrison Martland, who diagnosed her as having radium poisoning.

Mrs. La Porte was found to be suffering from a radium-induced osteogenic sarcoma of the pelvis. She was given deep x-ray therapy. For a few months, the therapy controlled the tumor. However, the sarcoma gradually grew and spread to other parts of her body.

In May 1931, Mrs. La Porte filed a lawsuit against the U.S. Radium Corporation. A month later, she died at the age of 31. With her death, her lawsuit ended. However, a year later, her husband filed a new lawsuit against the company for damages for her injuries and death.

Again, the U.S. Radium Corporation claimed that the statute of limitations had run out. In January 1933, Mr. La Porte's lawyer moved the lawsuit to equity court to stop the company from using that defense. La Porte's lawyer argued that the statute did not apply because the corporation had fraudulently concealed the existence of the disease, and that the company was negligent in protecting its workers from the hazardous element.[61]

The case, which became a battle among experts testifying for both sides, dragged on for years. Finally, in December 1935, the federal court ruled:

> ... it is difficult to understand how the defendant [the U.S. Radium Corporation] can be said to have perpetrated a fraud when it, a commercial enterprise, failed to draw inferences, which no doctor had ventured, to the effect that its industry was suicidal to its workers. ...
>
> It is apparent from the literature produced during this trial that prior to 1925, although there were some suggestions of hazard through the agency of radium, many hailed it as a great boon to humanity, and its internal uses by injection, inhalation, etc., were frequently advocated.
>
> There is no question but the defendant was utterly ignorant of the harmful effects attendant upon its factory process until 1924, when its attention was directed to an alleged case of radium necrosis suffered by one of its former employees.
>
> It is tempting in the light of the knowledge of today and the experience since 1920 to create the thought that the defendant must have been negligent in some way. Today, industrial methods which the defendant then employed would not be merely negligent but criminal. But it should be carefully noted that this case must be decided on the facts as they existed in the light of the knowledge of 1917 to 1920.
>
> Naturally there is no question as to where the sympathies of any human being would lie in a case of this sort. But a court has no power to adjust the law which has been enacted to meet the needs of a time when no such case as this could be foreseen. This is an extraordinary case even today. The statute of limitations was enacted for the purpose of protecting the public from fraud. Its ends were desirable and necessary, and in the infinite variety of cases that come before the courts that is still true. The responsibility in this case can only be laid to the tremendous progress made in science in the last four decades, for radium was unknown prior to 1898. The development of the law to meet such contingencies must of necessity lag behind their discovery. Only forward looking, intelligent legislation can protect future situations such as the one here present.
>
> The bill must be dismissed.[62]

With this ruling, all of the dial painters who had the disease but had not filed suit, as well as those who would eventually develop the disease in the future, were denied justice. As one author commented, the outcome of the case was a "demonstration of the merciless administration of law."[63]

Unfortunately, the tragedy would continue for decades as the former workers died slowly from the effects of the radium. Of the approximately 800 dial painters who worked at the Orange, New Jersey, radium plant from 1917 until 1924, a total of 43 women are known to have died from their exposure to the radioactive element.[64]

Once again, the world would be warned of the hazards of internal radioactivity, this time by the dial painters of a small Illinois town.

Chapter 7

The Ottawa Society of the Living Dead

Graveyards are for the old and tired, not for the young and life-loving. But in Ottawa graveyards ... there are too many young girls and mothers, victims of radium poisoning—the human discards of one of the city's industries.[1]
Mary Doty, 1936

The Radium Dial Company

The story of the Ottawa, Illinois, radium dial painters parallels that of the New Jersey workers. However, the Ottawa tragedy would take place several years later, and it would center on the Radium Dial Company.

The Radium Dial Company, a division of the Standard Chemical Company, first established a small dial-painting "art studio" in Chicago in 1918. Initially, the studio moved from one downtown location to another and for a time was located in the Marshall Field Annex building. While in Chicago, the studio began painting dials for the enormous Western Clock Manufacturing Company (Westclox). Soon Westclox became their largest customer.

Late in 1920, Westclox persuaded the dial company to move its studio 100 miles southwest to the city where their clock factory was located, Peru, Illinois (Figure 34). Within a short time, however, Westclox began to complain that its best employees were being attracted by the higher wages paid for dial painting. After only two years the studio moved again, this time to the small town of Ottawa, 15 miles away.

In September 1922, the studio occupied the former Ottawa Township High School. This large Victorian building, which had been vacant for several years, consisted of two stories and a basement. Most of the dial painting took place in one large room on the second floor that was formerly the school auditorium (Figure 35). The new studio

Figure 34. *Young dial painters at the Peru Radium Dial Company, 1922.*

employed approximately 100 workers, and almost all were dial painters.[2] Most were young women in their teens and early twenties.

Working conditions seemed ideal: rooms were well lit, the pay was excellent, and the work was enjoyable. One worker described the position of dial painter as "the elite job for the poor working girl."[3]

New employees were trained to use the glowing radioactive paint during a one-month trial period. They had to learn the tasks of mixing the paint, keeping it in suspension, and applying it evenly and neatly to each of the dial's numerals. At the end of the trial period, those who were considered good painters were offered a permanent position; the others were dismissed.

All of the dial painters were specifically taught to point or tip the brush in their mouths to obtain a fine point. To convince new employees that the glowing paint was "harmless," those who taught them would sometimes eat a quantity of paint from a spatula.[4] Workers also were told that the paint would help their complexions and make them beautiful.[5]

The practice of brush pointing varied greatly among the dial painters. Some workers infrequently tipped their brushes and ingested only a very small amount of the paint; others, however, frequently tipped them and virtually ate the paint.[6]

Figure 35. Dial painters at the Ottawa Radium Dial Company, 1924.

To sharpen their skills, the Ottawa dial painters were encouraged to practice painting at home by decorating the buttons and belts of their dresses. They also were encouraged to paint common household items such as key chains and lamp pulls. One woman, who would eventually die from radium poisoning, even painted her bedroom walls to make them glow in the dark.[7] When the supervisor was not looking, others painted their fingernails, eyebrows, and teeth, and streaked their hair with the glowing paint before going out on dates.[8]

New workers would start out painting large clock faces and hands such as those used on Big Ben alarm clocks. As they became more proficient, they would be given smaller alarm clocks, and eventually pocket watch dials, to paint. At the Ottawa studio, an exceptionally skilled worker could paint about 200 dials per day. Because the women were paid on a piecework basis, weekly earnings varied considerably. In 1925, for example, the weekly salary ranged from $17.00 to an occasional high of $42.00.[9] These were very desirable wages for young women in the 1920s, and the dial painters ranked with milliners in the upper 5% of female wage earners nationally.[10]

From its beginnings in 1918 until 1924, the Radium Dial Company would employ about 1,000 young women, with practically the entire personnel changing with each move the company had made.[11]

By 1925, the Ottawa studio had become the largest dial-painting establishment in the country, producing 4,300 dials daily, or approximately 1,075,000 per year. It painted all the radium dials used by Westclox, a company that produced more radium dial clocks and watches than any other manufacturer in the United States.[12]

Deaths and Denials

When the management of the Radium Dial Company first became aware in 1925 of the lawsuits filed by the New Jersey dial painters, they quickly took several steps. First, the company had all of their Ottawa workers physically and medically examined. X-rays and blood samples were taken, and the women were tested for the presence of radon in their breath, a positive breath test indicating the presence of radium in the body. Some of the painters were indeed found to be positive, but they were never told.[13]

Next, to ensure that production would not be disrupted if the Ottawa dial painters refused to work because of the bad publicity, the firm established a second studio in the nearby town of Streator, Illinois. According to a Department of Labor report of 1929 on radium poisoning: "Dial painters in other studios in different States became frightened through the publicity of the New Jersey cases. ... One of these [firms] was compelled to start an auxiliary studio in another town where less was known about it."[14] Both studios operated simultaneously for nine months (April–December 1925) until it was clear that the Ottawa workers would remain at their jobs.

Finally, in late 1925, after experimenting with various alternative methods of applying the radium paint, the company issued glass pens to the dial painters and banned them from pointing brushes in their mouths under penalty of dismissal. The pens would keep a permanent point but were very clumsy to handle, and a sponge had to be used to clean off the excess paint. Some workers, however, continued to use brushes and, without thinking, to point them.

Initially, the Ottawa dial painters were not alarmed by the news of the New Jersey radium poisoning victims. But as more and more headlines appeared, they became increasingly frightened, almost to the point of panic. "We were extremely happy in our association," said one dial painter, "until news of the poisoning and deaths from radium poisoning ... were printed. Then the girls became wild. There were meetings at the plant that bordered on riots."[15]

The situation reached a critical point during June 1928. That month, the dial painters were again tested for radon. Of the 67 dial painters tested, 10 were found to be positive, 13 were classified as being very suspicious, 11 were rated as suspicious, 12 were considered doubtful, and 21 were classified as negative.[16]

Again, the women were not informed as to the results of the tests despite their repeated requests. However, they were told by their supervisor: "My dear girls, if we were to give a medical report to you girls, there would be a riot in the place."[17]

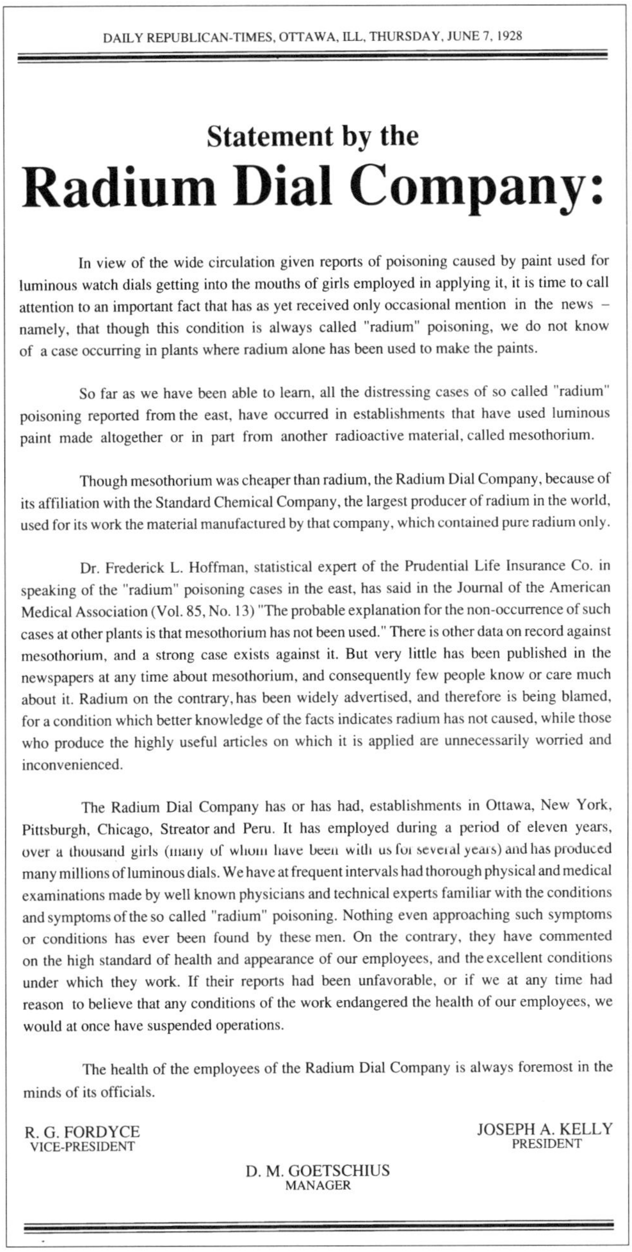

> DAILY REPUBLICAN-TIMES, OTTAWA, ILL, THURSDAY, JUNE 7, 1928
>
> ## Statement by the
> # Radium Dial Company:
>
> In view of the wide circulation given reports of poisoning caused by paint used for luminous watch dials getting into the mouths of girls employed in applying it, it is time to call attention to an important fact that has as yet received only occasional mention in the news – namely, that though this condition is always called "radium" poisoning, we do not know of a case occurring in plants where radium alone has been used to make the paints.
>
> So far as we have been able to learn, all the distressing cases of so called "radium" poisoning reported from the east, have occurred in establishments that have used luminous paint made altogether or in part from another radioactive material, called mesothorium.
>
> Though mesothorium was cheaper than radium, the Radium Dial Company, because of its affiliation with the Standard Chemical Company, the largest producer of radium in the world, used for its work the material manufactured by that company, which contained pure radium only.
>
> Dr. Frederick L. Hoffman, statistical expert of the Prudential Life Insurance Co. in speaking of the "radium" poisoning cases in the east, has said in the Journal of the American Medical Association (Vol. 85, No. 13) "The probable explanation for the non-occurrence of such cases at other plants is that mesothorium has not been used." There is other data on record against mesothorium, and a strong case exists against it. But very little has been published in the newspapers at any time about mesothorium, and consequently few people know or care much about it. Radium on the contrary, has been widely advertised, and therefore is being blamed, for a condition which better knowledge of the facts indicates radium has not caused, while those who produce the highly useful articles on which it is applied are unnecessarily worried and inconvenienced.
>
> The Radium Dial Company has or has had, establishments in Ottawa, New York, Pittsburgh, Chicago, Streator and Peru. It has employed during a period of eleven years, over a thousand girls (many of whom have been with us for several years) and has produced many millions of luminous dials. We have at frequent intervals had thorough physical and medical examinations made by well known physicians and technical experts familiar with the conditions and symptoms of the so called "radium" poisoning. Nothing even approaching such symptoms or conditions has ever been found by these men. On the contrary, they have commented on the high standard of health and appearance of our employees, and the excellent conditions under which they work. If their reports had been unfavorable, or if we at any time had reason to believe that any conditions of the work endangered the health of our employees, we would at once have suspended operations.
>
> The health of the employees of the Radium Dial Company is always foremost in the minds of its officials.
>
> R. G. FORDYCE
> VICE-PRESIDENT
>
> D. M. GOETSCHIUS
> MANAGER
>
> JOSEPH A. KELLY
> PRESIDENT

Figure 36. Facsimile of the "Statement by the Radium Dial Company."

At the same time, newspapers were filled with stories concerning the sensational trial and the out-of-court settlement of the lawsuits filed by the five New Jersey dial painters against the U.S. Radium Corporation.[18]

On June 7, 1928, three days after the settlement, the Radium Dial Company tried to counter the alarm and publicity by running a full-page advertisement in Ottawa's local newspaper (Figure 36); this advertisement also was posted in several rooms of the studio, and all the dial painters were advised to read it. The advertisement, entitled "Statement by the Radium Dial Company," bore the names of the firm's manager, vice president, and president.[19]

The "statement,"which is both highly deceptive and fraudulent, was clearly an attempt to lull the dial painters and the community into a false sense of security. It also was a thinly veiled threat indicating that the company would close the studio if the situation warranted it.

In the statement, Radium Dial blamed the use of mesothorium for radium poisoning and indicated that its paint never contained any of it, only "pure radium." In fact, the company had used paint activated by a mixture of radium and mesothorium for a number of years.[20] Also, as they were undoubtedly aware, the health effects of the two are virtually the same.

Radium Dial's statement cited an article by Frederick Hoffman, published in the *Journal of the American Medical Association* in September 1925.[21] They totally ignored a later, more definitive article by Dr. Harrison S. Martland in a December issue of the same journal. In the article, Martland warned "that when long-term radioactive substances are introduced into the body ... death may follow a long time after... ."[22]

Last, the Radium Dial Company clearly knew it was just a matter of time before some of their workers would develop radium poisoning. The company was aware that several of their employees tested positive in 1925 and again in 1928 for the presence of radium. A U.S. Department of Labor report described the situation in the following words:

> While all known cases of poisoning from radioactive substances were for a long time confined to the single New Jersey studio, the first one to start dial painting and probably the largest in the country until the appearance of the disease, similar cases finally came to light in other localities. This showed that the poisoning was not due to some conditions peculiar to this one studio, and it also created apprehension among the various firms who were engaged in the work.
>
> Some of these firms had, however, known that outbreaks [of radium poisoning] might be expected at any time, because the investigation conducted by different firms had disclosed radioactivity among the dial painters in some of the studios. The knowledge of these discoveries had, however, been carefully concealed by the firms, who feared disruption of their business if the facts became known. Even the victims had not been informed of their condition, nor the cause, through fear of panic among the workers.[23]

Regardless of its credibility, the Radium Dial Company's statement defused the situation. The Ottawa dial painters remained at their jobs, and few lawsuits were filed until years later.

The first Ottawa death from radium poisoning occurred in 1927. A young dial painter, Mary Ellen Cruse (1903–1927), had worked for the Radium Dial Company intermit-

Figure 37. Mary Ellen Cruse.

tently for almost three years (Figure 37). A year before her death she developed a hard ridge under her chin and complained of pains in her hands and legs. She began to tire easily and developed pains in her jaw. A week before her death she opened a small pimple on her face, and when it began to swell, a physician was consulted. Her face continued to swell, and her physician ordered her to enter the hospital at once. She died a few days later at the age of 24. Her death was misdiagnosed as "streptococcal septicemia, infection of face."[24]

A year later, in July 1928, Mary Ellen Cruse's parents filed suit with the Illinois Industrial Commission for $3,750 against the Radium Dial Company for the death of their daughter. After years of litigation, they would receive a settlement of $500.

Figure 38. Margaret Looney (left).

The next radium dial painter's death occurred in 1929. Margaret Looney (1905–1929) had first worked for the radium company in January 1923 (Figure 38). She continued working there for six years, until shortly before her death. By all accounts she was a very conscientious worker, even painting dials at home.

When the company conducted examinations of their employees in 1925, Margaret was found to be positive for the presence of radium. A year later, she had a tooth extracted, and her jaw was fractured. The following year, a second tooth was removed, and the socket did not heal promptly. When the company tested its employees in June of 1928, Margaret was again found to be positive for radium. Soon she began to suffer from weakness and anemia. She was sent to a university hospital clinic in Chicago, where the physicians did not diagnose the true cause of her illness but advised her to change employment. However, she continued to work at the studio, and her condition worsened. According to her mother: "You could see her slowly dying. There was nothing you could do. 'Well, mother,' she used to say, 'my time is nearly up.'"[25]

In early August 1929, Margaret became ill at work and was sent by the company supervisor to a local hospital. In her weakened condition from radium poisoning, she soon developed diphtheria and pneumonia. She was placed in isolation and died three days later at the age of 23.

A few months before Margaret's death, the U.S. Department of Labor, which was conducting a major study of radium poisoning in the nation's dial-painting industry, requested information about her from the Radium Dial Company. The department had learned (probably through the interviews it had conducted) that Margaret had tested positive for radium in both 1925 and 1928 and was aware of her dental problems, which were similar to those experienced by other radium poisoning victims. However, the company would provide only her starting date of employment and informed the department that she was still working for the firm. According to the labor department's report on the case, "Complete information was not obtainable, and the firm protests against calling the diseased condition radium poisoning, but it seems well indicated by the ... test."[26]

The labor department's request for information alarmed and forewarned the company. When Margaret Looney died in Ottawa, the management of Radium Dial sent a physician from Chicago to conduct an autopsy. The family agreed to the autopsy if their physician was also present, and a mutually agreeable time was set. However, the Chicago physician conducted the autopsy early, and when the family's doctor arrived at the specific time, he found that the autopsy had already been completed.[27]

The Chicago physician's autopsy report stated, "The teeth are in excellent condition. The gums appear normal. There is no ulceration of the gums, nor any evidence of any destructive bone changes in the upper or lower jaw." The report gave the cause of death as diphtheria and bronchopneumonia, as did the death certificate.[28]

Ms. Looney's obituary in the local newspaper read, "The young woman's physical condition for a time was puzzling. She was employed at Radium Dial studios, and there were rumors her condition was due to radium poisoning." The obituary continues, "... the doctors discounted this and gave the cause of death as anemia and diphtheria."[29]

The family believed that the death certificate was in error. They claimed compensation from the company through the Illinois Industrial Commission but never received a settlement.

Almost 50 years later, when Ms. Looney's body was exhumed and measured for radioactivity, her body was found to have an extremely high terminal body burden of radium. Also notable was the finding that although almost all of her skeleton was present, a few bones were missing. A section of her upper jawbone, for example, appeared to have been removed at her autopsy.

During 1930, two more radium-poisoning deaths occurred. Two former dial painters died from radium-induced sarcomas. One woman who died was first employed by the Radium Dial Company when she was only 13 years old. She worked for the company for six years and died at the age of 21. Her death certificate listed her cause of death as sarcoma of the retroperitoneal lymph glands; her jawbone had disintegrated before her death. The other woman worked at the studio until she was married and then died at the age of 26. Her cause of death was given as sarcoma of the hip.[30]

In 1931, the fifth victim, Pauline Kenton (1901–1931), died. Her death certificate was the first to identify radium poisoning as the underlying cause of death. Sarcoma of the pelvis was given as the secondary cause. She died in Maryland, and although her obituary appeared in the *New York Times*, no one in Ottawa is known to have seen it. However, the paper neither mentioned the city where she was employed, nor the name of the dial-painting firm for which she had worked. In addition, this woman appears to have been employed by the Radium Dial Company mainly in Peru, Illinois, and she may have worked at the Ottawa studio for only a short time.

The next death, which occurred in 1934, was apparently the first to be recognized in Ottawa as being due to radium poisoning. This death convinced a small but growing number of residents that they were confronting a new occupational disease; however, others, including the physicians in the community, remained skeptical.

Mary Robinson (1906–1934) was an employee of the Ottawa studio for three years, 1923–1925. After leaving the company, she soon became ill. For seven years, her illness became progressively worse until her death. At first, she developed terrible pains in her legs, which became swollen and misshapen. Later, the pain spread throughout her body. Several months before her death, her arm "burst open" and had to be amputated. According to a newspaper interview with Mary's mother:

> Mary's was the first case definitely called radium poisoning.... The doctors scoffed when I mentioned radium at first, but as soon as I read about those girls in New Jersey, I knew what was the matter with her. Then, just before they took her arm off, they sent a sliver of bone to a New York laboratory. They sent back word it was radium poisoning. The Ottawa doctors couldn't deny it then.

She goes on to state: "Mary's death certificate, signed by an Ottawa physician, gave the cause of death as "generalized sarcoma." However, the physician answered "no" to the question on the death certificate that asked, "Was disease in any way related to occupation of deceased?"[31]

Ottawa's physicians steadfastly refused to admit that radium poisoning was the cause of the women's illnesses and deaths until years later, when outside medical evidence conclusively proved otherwise.

Legal Battles

The year 1934 marked the beginning of a series of long and difficult legal battles by Ottawa's radium poisoning victims to obtain compensation. The battles would be fought in many courtrooms, and the ultimate outcome would be a hollow victory for the women.

Also during the year, the president of the Radium Dial Company, Joseph A. Kelly, Sr. (1884–1965), would be ousted by the other stockholders of the firm for allegedly attempting to defraud the company by making its stock worthless.[32] However, over a three-day period, Kelly and several associates would establish a new dial-painting company, Luminous Processes, Inc., a few blocks from the Ottawa Radium Dial studio. The two companies would compete for contracts with Westclox, and the new company would ultimately drive the Radium Dial Company out of business.

With the increasing number of deaths, and the refusal of local physicians to recognize the new occupational disease, a small group of former dial painters attempted to take the situation into their own hands. A member of the group contacted a Chicago physician, Dr. Charles Loeffler, and invited him to Ottawa. The group met with Dr. Loeffler at a local hotel, where he held an informal clinic. After concluding his examinations, he advised many of the women that they were indeed suffering from radium poisoning.

Soon after these examinations a number of the women signed a contract with a Chicago attorney, J. S. Cook, to bring suit against the company. No Ottawa lawyer would take their case because many of the town's "residents bitterly resented these women's charges as giving a 'black eye' to the community."[33]

On the advice of their lawyer, two women went to the studio representing the group to see the supervisor in order to legally notify the company that they were ill and were seeking compensation and medical care. The supervisor responded by saying that he didn't think there was anything wrong with the women. He also denied that radium poisoning existed, saying that "there was nothing to it at all."[34]

In July, seven lawsuits, each asking for $50,000 in damages, were filed in the Superior Court of Cook County against the Radium Dial Company. The suits charged that the company violated the Illinois Occupational Diseases Act and that the women were suffering from anemia, rarefaction of the bones, alveoli of the jaws, and other bone complications and disorders. The suits also indicated that the women had spent the greater part of their savings on medical care.

Specifically, all of the suits alleged that during the process of dial painting, particles of dust consisting largely of radium were thrown off, and the workers inhaled, swallowed, or otherwise took these particles into their systems, and that the company was in violation of the Act because it was careless and negligent in failing "to provide reasonable and approved devices, means or methods for the prevention of such occupational disease ... ."[35]

The first lawsuit filed, that of Mrs. Inez Corcoran Vallat (1907–1936), became the test case for the other suits. Mrs. Vallat had worked at Radium Dial for six years, starting with the company in 1923. After several years, she began to have muscular and joint aches and pains. After two years of constant pain, she went to the Mayo Clinic in 1929, at the age of 22. Doctors there diagnosed her as having "late radium effects on her bones." She soon developed problems with her jaw and returned to the clinic in 1930, where an infected molar was removed. In the following years, she developed other ailments. According to one newspaper account: "For five years her hips were locked so that she could move neither backward nor forward, and for the last three years one side of her face drained constantly."[36]

During the trial, the lawyers representing the radium company argued that Mrs. Vallat's health problems had not occurred during her period of employment; that the two-year statute of limitations had expired before she filed suit; and that poison was not covered by the Illinois Occupational Diseases Act.

The company won the case in two lower courts. However, Mrs. Vallat appealed to the Illinois Supreme Court. Now, Radium Dial's lawyers argued that the state's occupational diseases act itself was unconstitutional. And on April 17, 1935, the court ruled that the act was indeed unconstitutional because its provisions were "so vague and indefinite that no person by reading the enactment can know with reasonable certainty what rights it confers and what obligations and duties it imposes upon employers. ..."[37]

With this decision, all of the lawsuits filed by the Ottawa dial painters were dismissed. Their attorney withdrew from the case, and the women were unable to get another lawyer to represent them. Ten months later, at the age of 29, Mrs. Vallat died an agonizing death from a hemorrhaging sarcoma of the neck.[38]

Newspapers called the decision a travesty. For example, the Chicago *Daily Times* termed it an "almost unbelievable miscarriage of justice, that now has added its disappointments to the doomed women's burden." The paper went on to say: "No man was at fault, no lawyer, no court. Simply, the law itself—specifically, Illinois' antiquated, insufficient 'Occupational Disease Act.'"[39]

Some of the women, who had also filed claims in early 1935 with the Illinois Industrial Commission, were still hopeful that a loophole somewhere in the law would permit them some redress for their suffering. In March 1936, the Illinois legislature enacted a new Workmen's Occupational Diseases Act which became effective October 1, 1936. The women's claims would now be heard under the new law.

For a while it appeared that no lawyer would represent the dial painters before the Industrial Commission. The women had little money, and the radium company had left the state and no longer had any assets in Illinois (other than a $10,000 bond they had posted with the commission). Thus, any settlement would be limited to $10,000 no matter what the commission might "award."

Their former lawyer refused to accept the case. He suggested they contact someone who might take their case for charity, such as the world famous Chicago attorney Clarence Darrow (1857–1938). They contacted Darrow, and although he was sympathetic, he declined to take their case, saying he was too old. Darrow did, however, refer the women to another Chicago lawyer, Leonard J. Grossman (1891–1956), who specialized in workmen's compensation cases.[40]

In July 1937, Grossman, a former Chicago ward alderman and assistant corporation counsel for the City of Chicago, accepted their case. It was clear to Grossman that the case would be difficult to win and that even if he were successful, he would receive little for his efforts. He was, however, entirely sympathetic to the women's plight (Figure 39).

Figure 39. Mrs. Donohue (center) and other members of the "Living Dead" signing a petition for Leonard Grossman to take their case, 1937.

The same month the new president of the Radium Dial Company, William Ganley, who was interviewed in New York City concerning the Ottawa cases, remarked:

> These women's claims are invalid and illegitimate. We beat several suits in the lower courts in Illinois, and the verdicts in our favor were upheld by the state Supreme Court.
>
> We don't feel any legitimate claims have been filed against us. A lot of these women were with us only a few months and never did any direct work with the radioactive salts we use in our process. Practically all of them have been out of our employ for many years.
>
> I can't recall a single actual victim of this so-called 'radium poisoning' in our Ottawa plant.[41]

A Miracle or a Happy Death

Attorney Grossman picked one of the dial painters' lawsuits as a test case. He chose that of Mrs. Catherine Wolfe Donohue (1903–1938). Her case would quickly become the most famous of all of the Ottawa radium dial painters.

Figure 40. Mrs. Catherine Wolfe Donohue.

Mrs. Donohue started working for the Radium Dial Company in 1922 at the age of 19 (Figure 40). She worked for the firm for almost nine years until 1931 (eight years as a dial painter and one year as a stockroom clerk). After working for 2 years, she began to develop a pain in her left ankle. The pain gradually grew worse and spread into her hip. Her hipbone became rigid, and she soon began to limp. Mrs. Donohue also began to have fainting spells that became increasingly frequent. Later, she developed soreness in her jawbone and teeth.

After reading about the New Jersey radium poisoning victims, she went to several physicians in Ottawa to be examined. "Two of them told her there was a destructive force in her teeth, but all of them were unfamiliar with radium poisoning cases and would not certify as to the cause of her condition."[42]

In 1931, when the radium company again conducted medical examinations of its employees, Mrs. Donohue was excluded from the tests. Despite her repeated requests, her supervisor refused to let her be examined.

By August 1931, Mrs. Donohue's health had so deteriorated that the managers of the company decided to dismiss her because the other workers were becoming alarmed. Both the president of the company, Joseph A. Kelly, Sr., and the vice president, Rufus Fordyce, came to Ottawa and told Mrs. Donohue that her "limping condition was causing talk, and it wasn't giving a very good impression of the company and that they felt it was their duty to let [her] go."[43]

In 1934, pieces of Mrs. Donohue's jawbone began to fall out. She went to a local dentist for treatment, but he was of little help. In March of the same year, she was examined in Ottawa by Dr. Loeffler at one of his informal clinics. Loeffler saw her several times and took blood samples for study. Soon he suggested that she come to Chicago to have x-rays taken of her jaw and hip, referring her to several physicians including Dr. Walter W. Dalitsch (1898–1971), an oral surgeon on the faculty of the University of Illinois College of Medicine and College of Dentistry.

Mrs. Donohue came to Chicago and saw Dr. Dalitsch. X-rays were taken which revealed that Mrs. Donohue was suffering from a malignancy of the hip, necrosis of the jaw, and radium poisoning.

Upon hearing of her diagnosis, Dr. Loeffler immediately telephoned the vice president of the radium company to express his concern. He "… told him from the cases I had seen, I thought it would be wise to investigate all the [other] cases, and ascertain whether radium poisoning existed in those cases, and if so, to try to prevent the rapid destruction in the cases I have [already] examined."[44] The vice president, however, refused to do anything.

Mrs. Donohue and the other dial painters filed lawsuits against the Radium Dial Company in 1934. Her suit, like the others, was dismissed after the Illinois Supreme Court's rejection of the Vallat case. She also filed a claim in 1935 with the Illinois Industrial Commission. After several delays, her hearing took place in Ottawa on February 10, 1938.

Mrs. Donohue's lawyer, Leonard Grossman, sought to prove that the Radium Dial Company had misrepresented and fraudulently concealed the industrial hazard of radium. He cited the "fake" company newspaper statement of June 7, 1928, as an example, saying that it was "false and misleading." He charged that the firm was grossly negligent in failing to warn its employees of the dangers of radium in the face of the growing evidence and in not providing protective equipment and work clothing. Grossman also charged that the company was careless in permitting employees to eat in the room in which they painted.

Radium Dial's legal defense was that radium was not a poison. Ironically, in the Vallat case the company had previously maintained that it was a poison, at a time when Illinois' old occupational disease act did not cover injuries from poisons. Radium Dial's lawyers also argued that the statute of limitations had expired.

Leonard Grossman and the attorney representing the radium company agreed that the final decision of the case would govern the claims of all of the other dial painters. They both knew that the case would be appealed and that the final ruling would be 18 months away.[45]

By the time of the hearing, Mrs. Donohue's condition had greatly worsened, and she weighed only 71 of her original 125 pounds. Describing her family and her poor state of health, one relative stated, "Catherine was full of radium and dying by inches ... she suffered agonies, and [her husband] Tom was nearly bankrupt buying medicines to try to relieve Catherine."[46]

During this time the plight of the Ottawa women gained press attention. Chicago newspapers referred to them as Ottawa's "Doomed Women," "The Living Dead," and "The Society of the Living Dead."

The industrial commission hearing took place in a packed courtroom. Mrs. Donohue was called first to testify. Looking old (despite the fact she was only 35), worn, and frail, she had to be carried into the courtroom. Speaking with difficulty in a soft, almost inaudible voice, she described how she was taught to point the brush between her lips, how the company had dismissed her because of her limping, and how she was unable to attend church services because of her health problems. Mrs. Donohue then went on to describe in detail how her health had deteriorated during the past several years, and she presented to the court a small jewelry box containing several pieces of her jawbone that had fallen out.

The next witness to testify was Dr. Dalitsch. He explained the basis for his diagnosis that Mrs. Donohue was suffering from radium poisoning. When asked whether Mrs. Donohue's condition was fatal, she fainted and collapsed before he could answer. After she was carried out of the courtroom, Dalitsch continued, saying that Mrs. Donohue's illness was indeed fatal and that she would live only "from months to a year or two," depending upon the treatment.[47]

Other witnesses testified, including Dr. Loeffler and a second physician from the University of Illinois and Mount Sinai Hospital of Chicago. They both confirmed that Mrs. Donohue was suffering from radium poisoning.

Because of the poor state of her health, the hearing was continued the next day at Mrs. Donohue's home, where she was cross-examined as she lay on a sofa in her front room (Figure 41). There, she demonstrated how she pointed the paintbrush in her mouth. Other witnesses were called, including Mrs. Charlotte Purcell (1906–1988), whose arm was amputated in 1934, before her condition was diagnosed as being due to radium poisoning. Mrs. Purcell also demonstrated how she was taught to point the brush in her mouth.

On April 5, 1938, the industrial commission ruled that the Radium Dial Company was liable for Mrs. Donohue's disability and directed payment to her of $3,470 as an award (calculated at $11.00 per week from April 1934, when her disability started, to June 1940), a life pension of $277.60 annually beginning on June 1, 1940, and $2,500 for past medical expenses.[48]

Figure 41. The second day of the hearing held at the Donohue home. Mrs. Purcell can be seen behind Mr. Grossman.

Mrs. Donohue felt that the settlement was wonderful and indicated that she had never dreamed of such an early decision. But, the very next day the Radium Dial Company's lawyers filed an appeal. The company would continue to challenge the settlement, and Mrs. Donohue would die without receiving a cent.

In June 1938, Mrs. Donohue wrote a deeply moving letter to Rev. James Keane of Our Lady of Sorrows Roman Catholic Church in Chicago, asking that prayers be said in her behalf. The letter read in part:

> The doctors tell me I will die, but I mustn't. I have too much to live for—a husband who loves me and two children I adore. But, the doctors say, radium poisoning is eating away my bones and shrinking my flesh to the point where medical science has given me up as 'one of the living dead.'
>
> They say nothing can save me—nothing but a miracle. And that's what I want—a miracle. ... But if that is not God's will, perhaps your prayers will obtain for me the blessing of a happy death.[49]

Mrs. Donohue received national attention and sympathy when Rev. Keane read her "inspiring but most pathetic letter" over the air during his weekly broadcast on Chicago radio station WCFL.[50] Her letter also was widely quoted in newspapers. Churches across the country asked that prayers be said for her, and she received thousands of cards and letters.

On July 27, 1938, Catherine Wolfe Donohue died at the age of 35. Her funeral was attended by all of the other members of "The Society of the Living Dead" and hundreds of other mourners.

Her death was certified as "radium poisoning," and the inquest jury cited the Radium Dial Company as "the only industrial plant she ever worked in." The coroner's physician went beyond the findings on the death certificate, stating that a bone sarcoma was the major cause of death.[51]

Shortly after Mrs. Donohue's death, Mr. Grossman pressed the radium company to pay at least half of the settlement to her estate. He indicated that if they refused, he would attempt to have the officers of the company held criminally responsible.[52]

However, the Radium Dial Company continued to delay by appealing. After losing an earlier request that the Illinois Industrial Commission overturn Mrs. Donohue's settlement, the firm's lawyers attempted to file an appeal with the clerk of the La Salle County Circuit Court, but they failed to post a statutorily required appeal bond. That court would not accept the appeal without the bond. Next, the company challenged the requirement of an appeal bond to the circuit court, alleging that the company was unable to obtain a bond or give security and thus that the required appeal bond violated its constitutional right to judicial review of this and other similar cases that were pending against it—all of which the radium company claimed were void because the industrial commission lacked jurisdiction to enter an award. The circuit court, however, rejected the company's argument.

The Radium Dial Company's lawyers next petitioned the Illinois Supreme Court for a decision on the narrow issue of the constitutionality of the appeal bond requirement. On June 15, 1939, the court rejected the company's petition, upholding the requirement. This decision was then appealed to the U.S. Supreme Court, which on October 9, 1939, also rejected the appeal "for want of a substantial Federal question."[53] This was the seventh time that Catherine Donohue's case had been "won."[54]

Although the details are unclear, Mrs. Donohue's estate probably received the original settlement of approximately $6,000, leaving the remainder of the $10,000 bond spread among the other dial painters suing for compensation. Perhaps they received a few hundred dollars each for their years of litigation and suffering.

Unfortunately, the tragedy would not end with the death of Mrs. Donohue, but would continue to haunt Ottawa and its workers for many decades. In total, 35 women are known to have died from radium poisoning.[55]

The deaths of the dial painters would put an end to the practice of brush pointing in the industry. However, radium was still being widely used in various medicines and tonics.

Chapter 8
The National Radium Scandal

Radium has been tried for the treatment of almost all the ills to which the human body is heir. ... Like the x-ray, radium therapy has been passing through the shadows of illiterate misconceptions and unfortunately has been handled at times by the unscrupulous.[1]
Dr. Everett S. Lain, 1922

Uses and Abuses

Despite the tragic illnesses and deaths of the radium dial workers, the public continued to drink radium tonics, many physicians continued to give their patients radium injections and medicines, and quacks and frauds continued to peddle various radium nostrums. Radium was just too entrenched, too economically profitable, and too miraculous an element to possibly be considered harmful.

Beginning in 1925, however, things would slowly begin to change. At the end of that year, Dr. Martland published his pioneering article, "Some Unrecognized Dangers in the Use and Handling of Radioactive Substances," in the *Journal of the American Medical Association*. In the article, he clearly warned the medical profession of the dangers of using radium internally. He wrote:

> From our experience, it would appear that the intravenous injections of long lived radioactive elements or the internal administration of radium, mesothorium or radiothorium is highly dangerous on account of the late harmful effects. It is not warranted in any medical condition, as none of the known radioactive substances produce any specific or curative results. ... The value of radium waters is questionable, since most of them at the time they are taken contain little or no radioactivity. Should the waters contain radium or mesothorium in solution, their use would be distinctly dangerous on account of late cumulative effects.[2]

Initially, Martland's warning would go unheeded. And he would be severely criticized, laughed at, ridiculed, insulted, and threatened by the producers of radium medicines and radioactive waters, physicians, scientists, and quacks.[3]

Despite his warning, in 1927, three physicians at the prestigious Mayo Clinic conducted experiments with radium injections. The doctors gave the injections to a total of 50 patients to see how effective radium was in lowering blood pressure, and in controlling pain.[4]

As late as 1931, several physicians, working with a professor of chemistry, also conducted experiments with radium injections. They gave 32 psychiatric patients at the Elgin State Hospital in Illinois the injections over a six-month period to determine their retention of radium and to observe any clinical changes. As a result, years later, several of the patients would suffer and die from radium poisoning.[5] It would take nothing less than a shocking national scandal to finally put an end to the widespread internal use of radium.

The National Radium Scandal

Because of radium's enormous popularity, many quacks and swindlers developed and sold radium health products, medicines, and tonics. Most of them, however, did not contain any of the costly radioactive element, and they were harmless. A few of them, though, actually did contain radium, and some in dangerously high doses.

The most notorious radium swindler of them all was William J. A. Bailey (1884–1949). Bailey was born in Boston to a large family. His father, a cook, died when he was very young, and his mother had to raise the children alone on a very limited budget.

One of Bailey's older brothers, Frederick, went to Harvard University and became a highly respected Boston physician. Following in his brother's footsteps, Bailey also attended Harvard. However, after three semesters he ran out of money, and he left the university without receiving a degree. Later, he would falsely claim that he graduated, and that he had gone on to receive a doctorate from the University of Vienna. After leaving Harvard, Bailey drifted from one job to another and traveled extensively. Eventually, he resided in New York City (Figure 42).

Bailey's first scam occurred around 1914. At that time, he established an automobile manufacturing company with the lofty title of the Carnegie Engineering Corporation. The corporation, which claimed to have the backing of the giant Carnegie Steel Company, advertised that it would produce a new inexpensive automobile, which would be sold for $595.00. The cars, which would be available on September 1, 1915, would be constructed at the corporation's manufacturing plant in Kalamazoo, Michigan, and shipped to Pittsburgh for delivery.

Bailey advertised his automobile throughout the United States, Europe, Africa, Australia, and South America. The advertisements indicated that advance orders for the car were being accepted for a $50 deposit. In a short time, thousands of orders poured in, totaling $750,000.

Eventually, several people became suspicious and filed complaints against the corporation with the U.S. District Attorney of New York City. When the government

Figure 42. William J. A. Bailey.

authorities conducted an investigation, they found that the corporation's manufacturing plant consisted of an abandoned sawmill, whose only furnishings were a single box of tools. They also found that although the corporation claimed to have one million dollars in capital, its only assets were a suite of offices, some stationery, and enough office equipment to keep three secretaries busy to collect orders.[6]

Bailey was charged with using the mails to defraud. He pleaded guilty and was sent to prison. After serving 30 days of his sentence, however, he was paroled in custody of his brother, Frederick.

Apparently Bailey's prison time taught him an important lesson for he never again would conduct corporate and financial fraud. Instead, he would move into the much less regulated area of medical fraud and quackery.

Around 1917, Bailey, along with another individual, developed and began selling a supposed aphrodisiac called "Las-I-Go for Superb Manhood." When the pills were analyzed, their active ingredient was found to be strychnine. And in 1918, Bailey was charged with false and fraudulent advertising and fined $200, plus court costs.[7]

In the early 1920s, Bailey "discovered" radium. His fascination with the radioactive element may have resulted from his employment at the U.S. Radium Corporation, where he worked for a brief time. For more than a decade, Bailey would develop and market one radium medicine or health product after another.

Bailey's first "radium medicine" was "Arium," which he claimed was radium in tablet form. Bailey widely advertised Arium, claiming that it could cure a number of diseases. However, when the Food and Drug Administration analyzed the tablets, they found that they did not contain any of the element at all. In time, a court declared that the claims made by the medicine were false and fraudulent, and it ordered the medicine seized and destroyed.

One of Bailey's radium "health" products was the "Radiendocrinator." The Radiendocrinator was a gold-plated radium disk three-eighths of an inch thick and two inches by three inches in size. Bailey claimed the device would benefit various glands of the body, such as the thyroid and the adrenal glands. However, the device's primary use was for "sexual rejuvenation." Supposedly, if a woman wore the Radiendocrinator around her waist it would rejuvenate her ovaries, and if a man wore it under his scrotum it would rejuvenate his testes and restore and enhance his sex drive. In addition, Bailey claimed the device would cure numerous medical conditions such as acidosis, baggy eyes, constipation, dry scalp, fatigue, flatulence, obesity, poor memory, and pimples. Originally, the Radiendocrinator—which actually did contain a large quantity of radium—sold for a pricey $1,000. Later, as demand declined, it was reduced to $150.[8]

Early in 1925, Bailey moved to East Orange, New Jersey, where he established a company called the Bailey Radium Laboratories. Here, he would produce his most infamous radium medicine. That year, Bailey purchased concentrated radium and mesothorium fluid from the U.S. Radium Corporation. He then diluted the fluid with distilled water and produced a highly radioactive tonic (Figure 43), which he called "Radithor."

Bailey packaged Radithor in half-ounce bottles and sold them by the case. A case consisted of thirty bottles, a full month's supply for the average condition. A case, which contained about $7.00 worth of radium, retailed for $30.00. Bailey, however, offered physicians a professional discount of $5.00.

Every case of Radithor contained a printed guarantee that declared: Every bottle of the tonic contained genuine radium and mesothorium in triple-distilled water; the strength of the solution in each bottle was the same; the tonic was produced under strictly sanitary conditions in thoroughly sterilized bottles; the tonic's physiological results were due entirely to the action of the radioactive element it contained; and, lastly, that Radithor was "harmless in every respect."[9] To further bolster his guarantee, Bailey offered

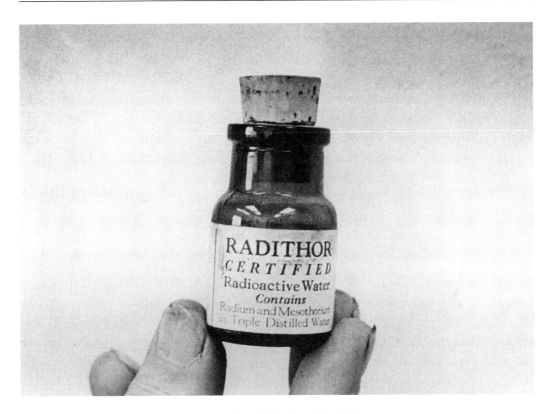

Figure 43. A bottle of Radithor.

to pay $1,000 to anyone who could prove that each and every bottle of Radithor did not contain a definite amount of both radium and mesothorium.

Like his other radium medicines and health products, Bailey widely and lavishly advertised Radithor. He called his tonic "the outstanding achievement in the application of radioactive rays" and "the climax of thirty years of toil by hundreds of scientists who labored with invisible rays that the cause of humanity might be served." Bailey also claimed that Radithor would not only cure or improve more than 160 medical conditions or symptoms, but also that it was a metabolic stimulant and a powerful aphrodisiac.[10]

Shortly after Bailey started selling Radithor, however, the deaths of the New Jersey radium dial painters began receiving publicity. Bailey reacted to the publicity by vigorously defending the U.S. Radium Corporation and the use of radium as an internal medicine. Bailey wrote countless letters to newspaper editors declaring that there was absolutely no proof that radium was responsible for the deaths of the dial painters. He pointed out that Marie Curie, who had worked for years with high concentrations of radium, was still alive and supposedly in excellent health. And he argued that, "If she has lived all these years around tremendous amounts of radium it would seem that persons handling minute amounts in painting watch dials should not die so readily."[11]

Bailey belittled the physicians who reported the new disease of radium poisoning, writing:

> It is to be regretted that physicians with no experience whatever in the field of internal radium therapy should make such ridiculous statements as have appeared regarding a so-called 'new radium disease.' It is passing strange that nothing like this has cropped out in the case of hundreds of thousands of people who have visited the spas of America and Europe for years or who have been treated at the radium clinics and in private practice for many years in this country and abroad with far larger doses than one could get from a carload of watch dials. It is a pity that the public has been turned against this splendid curative agency by the unfounded statements that have been appearing regarding radium deaths from luminous paint on watch dials.[12]

Last, Bailey wrote that in order to prove conclusively that radium was harmless he would personally take in one dose all of the radium used on all of the watch dials produced in a month at the U.S. Radium plant, or, for that matter, at any other radium dial plant. It is not clear whether Bailey actually kept his promise or not, although his bones would later be found to be highly radioactive indicating that he had indeed drunk Radithor.[13]

Eventually, as the publicity lessened, sales of Radithor increased. Bailey would sell his radioactive tonic through drug stores and physicians, and directly to thousands of consumers. From 1925 until 1929, he would sell an estimated 400,000–500,000 bottles.[14] Sales of the expensive tonic, however, would quickly come to an end with the onset of the Great Depression.

One of the physicians who purchased Radithor from Bailey was Dr. Charles C. Moyar (1881–1943) of Pittsburgh. Moyar, who received his medical degree from Jefferson Medical College of Philadelphia in 1905, was a well-known physiotherapist and internist.

Dr. Moyar clearly believed in the supposed medical benefits of Radithor; he drank it himself, and he prescribed it to his patients. Moyar also found that Radithor could be a highly profitable medicine: He purchased it for $25 a case, and then turned around and sold it to his wealthy patients for $500 a case.

Two of Dr. Moyar's wealthiest patients were Eben M. Byers and Mary Jennings Hill. Moyar would prescribe and supply them with Radithor; as a result both would eventually suffer and die horrible deaths from radium poisoning. Moyar himself would eventually die from a radium-induced sarcoma.

Eben MacBurney Byers (1880–1932) was an internationally known industrialist, sportsman, and playboy (Figure 44). Byers was born in Pittsburgh to a wealthy family. His father owned an iron-pipe manufacturing plant.

Byers received his education at an exclusive New Hampshire preparatory school and Yale University. After graduating from Yale, he entered the family's iron business. In a

Figure 44. Eben M. Byers.

short time, he became president and director of the company. After the death of his older brother, he also became president of the A. M. Byers Company of New York City. In addition, Byers served as a director of the Bank of Pittsburgh, the Pennsylvania and Lake Erie Dock Company, and the Bessemer Coke Company.

Byers was an avid sportsman and athlete. He was an expert trapshooter and tennis player. He enjoyed racing horses and owned a number of stables in the United States and England. Byers also was an excellent golfer, winning the U.S. national amateur golf championship in 1906.[15]

Byers first began drinking Radithor in December 1927. While he was returning from a Harvard-Yale football game, he accidentally fell from his railroad sleeping car berth and fractured his arm. Byers saw several physicians concerning his injury, and he also complained of generally feeling run-down. One physician, Dr. Moyar, advised him to regularly drink Radithor, saying that it would relieve his pain and that it was "a very good thing."[16]

At first the radioactive tonic seemed to invigorate and restore Byers' health, and he enthusiastically recommended it to his friends and business associates. Byers was so enthusiastic about the tonic that he gave bottles of it as gifts. He even gave it to one of his race horses.

Byers began drinking two to three bottles of Radithor a day. He continued to drink the tonic for about three years. In total, he would consume several thousand bottles of it, perhaps three times the lethal intake, if ingested all at once.[17]

Eben Byers' close friend, Mrs. Mary Jennings Hill (1887–1931), also drank Radithor. Like Byers, she too was an avid sportswoman and athlete and was very enthusiastic about the tonic. She would drink it almost exclusively for five years, consuming about 3,000 bottles.

Around 1929, Mrs. Hill began suffering from an unstoppable nasal drip, indicating the disintegration of her facial bones. She also became severely anemic and developed jaw necrosis. And during the last two years of her life she was bedridden. In October 1931, she died from radium poisoning at the age of 44.[18]

Eben Byers' first health problems started around 1930. At that time, he seemed to have lost the toned-up feeling he had experienced earlier. He began losing weight and suffering from severe headaches. He also developed pain in his jaw, and several of his teeth fell out.

At the end of the year, Byers went to a dentist in New York City who pulled a tooth. The tooth's socket, however, refused to heal and pus continued to come out. After several months, Byers' condition worsened, and he went to another dentist while he was in Palm Beach, Florida.

More of Byers' teeth were removed as well as parts of his jaw, which were found to be necrotic. However, the area again refused to heal and Byers began to hemorrhage. He was given another operation, but his condition continued to worsen. Finally, the Florida dentist recommended that Byers see an oral surgeon.

Byers left Florida and returned to New York City where he saw his personal physician. His doctor diagnosed him as suffering from a bad case of sinusitis and admitted him to a hospital. Byers was treated and released.

As his health continued to fail, Byers was hospitalized again. Byers' physicians were puzzled. His bones seemed to be disintegrating. X-rays were taken, and when the radiologist examined them, he recognized the signs of radium upon Byers' bones. Frederick B. Flinn of the Columbia University School of Public Health, who had examined many of the radium dial painters, was consulted. Flinn examined the x-rays and conducted several tests for the presence of radioactivity, and confirmed that Byers was indeed suffering from radium poisoning.

Byers' doctors, with the help of Flinn, tried various methods to remove the radium from his body. Despite their heroic efforts, there was no hope.

As Byers was slowly dying, the Federal Trade Commission began an investigation of Bailey and Radithor. In February 1930, the commission had filed a complaint against Bailey, accusing him of making false and fraudulent statements about the curative properties of Radithor and of misusing the term "laboratories" in his corporate name.[19]

To gather information concerning the complaint, the commission called Byers to testify. Byers, however, was so ill that he could not travel, and the commission had to send an attorney to his Southampton, Long Island, mansion to obtain his deposition.

When the attorney saw Byers in September 1931, he was shocked at his appearance. He would later comment:

> A more gruesome experience in a more gorgeous setting would be hard to imagine. We went up to Southampton, where Byers had a magnificent home. There we discovered him in a condition which beggars description.
>
> Young in years and mentally alert, he could hardly speak. His head was swathed in bandages. He had undergone two successive operations in which his whole upper jaw, excepting two front teeth, and most of his lower jaw had been removed. All the remaining bone tissue of his body was slowly disintegrating, and holes were actually forming in his skull.[20]

Although Bailey had already stopped producing Radithor by the end of 1929, in December 1931, the Federal Trade Commission issued a cease and desist order against him and his company.

In the meantime, Byers' health worsened, and early in 1932, he entered a hospital for the last time. His necrosis continued to spread and more of his jaw and parts of his skull were removed. Toward the end of his life, Byers developed a brain abscess, meningitis, and finally pneumonia. On March 30, 1932, Eben MacBurney Byers died at the age of 51.[21]

Byers' death from radium poisoning caused a national scandal. Newspapers and magazines across the nation reported his horrifying death. The media hounded Bailey and Dr. Moyar to get their sides of the story; however, both denied any responsibility. The media also repeatedly interviewed Dr. Harrison S. Martland and Frederick Flinn, both of whom issued strong warnings against the use of radium water and other radioactive products.

Government agencies reacted to Byers' death by starting several independent investigations. The Federal Trade Commission reopened its hearings on Radithor. Officials of the commission vowed to investigate other radium "cure" products. The Food and Drug Administration issued more warnings about radium waters and other radioactive products and medicines, and it asked for additional powers to better regulate. Finally, city and state health departments sent out inspectors to canvass drug distributors and drug stores to see whether they were selling radioactive medicines and products in violation of sanitary codes.[22]

The American Medical Association reacted by removing radium for internal use, in any form, from its list of "New and Nonofficial Medical Remedies." The association's Bureau of Investigation condemned radium quackery and published a stinging three-page exposé in its journal on the use of radium as a "patent medicine" and the methods and activities of Bailey.[23]

Byers' death thoroughly shocked and frightened the public. With his death the radium water, patent medicine, and health product industry collapsed. Several drug companies that produced radium products went bankrupt, and scores of sanatoriums that advertised their radium waters closed.[24]

Byers' death also would lead to important research on the health effects of radium. His death along with those of the dial workers would provide a vital warning to World War II's most important military project.

Chapter 9

Safety Standards and the Atomic Bomb

When I visited ... Oak Ridge early this month I was greatly concerned with the health hazards involved in the handling of plutonium. Today I wrote ... a letter emphasizing the need for new procedures. ... I said it might be necessary to institute protective measures of the type used in the radium dial industry... .[1]

Glenn T. Seaborg, 1944

The Gathering Storm

During the Great Depression of the 1930s, sales of radium dial clocks and watches sharply declined. As a result, radium application plants greatly cut back their production, and only a very small number of women were employed as dial painters.

With the outbreak of war in Europe in 1939, however, America began to prepare for the possibility that it might be drawn into the conflict, and the military demand for radium dials once again began to increase. Airplanes, for example, were now larger and more sophisticated than they were in the First World War, and they required a greater number of luminous instruments with higher concentrations of radium.[2]

To meet the growing demand, existing radium application plants expanded, and many new ones opened across the country. These factories needed to hire hundreds of new dial painters, despite the fact that it "was the most feared occupation among many young women."[3]

To ensure the new dial painters' safety, and to avoid repeating the tragedy of the past, occupational protection standards needed to be developed. The U.S. Navy would take the lead in pressing for such standards. To develop them, they would contact a young physics professor, Robley D. Evans, for help.

Robley Evans (1907–1995) first became involved in studying the health effects of radium by accident. After several years of study, however, he would become the world's leading authority on the subject (Figure 45). His friends would label him "Mr. Radium."

Figure 45. *Robley D. Evans and Nancy Caldwell demonstrating how a breath sample is taken to determine the amount of radium in the body, MIT, late 1950s.*

The accident that started Evans' long and illustrious career was the widely publicized death of the industrialist and sportsman Eben Byers. When Byers died from radium poisoning in 1932, his death shocked California health authorities, and they were determined not to have another, similar scandal. Accordingly, a Los Angeles County public health officer visited Professor Robert A. Millikan (1868–1953) at the California Institute of Technology (Caltech) about the problem. Millikan, who was a Nobel laureate in physics and president of Caltech, also was Evans' dissertation advisor.

Because Evans was completing his doctorate degree, which dealt with measuring the amount of radium in rocks, and because he was the only physicist at the time working in the field of radioactivity at Caltech, Millikan brought the health officer to see him.

Millikan introduced him to Evans and sternly instructed the health officer to "do what this young man tells you to."[4] The Nobel laureate then quickly turned and walked away.

Over the next several months, Evans reviewed the existing literature on the health effects of radium and became increasingly interested in the subject. In 1933, he wrote a comprehensive review of radium poisoning, which was published in the *American Journal of Public Health*.[5]

That same year, Evans began working with a Los Angeles physician who was treating an East Coast dial painter who was suffering from radium poisoning. The two tried to remove the radium from the woman: the doctor gave her various diets and medications, and Evans measured any changes in the amount of radium being shed from her body.

In 1934, Evans moved from California to join the physics faculty of the Massachusetts Institute of Technology (MIT). During his first year at the institute, he developed an important method of detecting and measuring the amount of radium contained within the human body, or radium body burden. He determined the burden based on the amount of radon in the breath and the amount of gamma rays (radiation similar to x-rays) emitted from the body.

This unique method eventually would be known as whole-body counting.[6] Evans later would verify his estimates of the body burdens taken during life by exhuming several of the dial painters and measuring the actual amount of radium within them.

In 1936, the U.S. Food and Drug Administration sent a representative to MIT to ask Evans to conduct research for them. The government agency was having problems with the continuing sales of medicines and cosmetics that contained radium. They asked Evans to establish some quantitative basis for determining safe amounts of radium in commercial products. At the time, there was not a sufficient number of radium poisoning victims who had been studied to suggest a permissible body burden of radium, or the maximum amount acceptable without the likelihood of bodily damage.

Evans tried to address the problem by conducting a series of animal experiments. He and several colleagues fed radium-dial compounds to rats over a 4-year period. Eventually, the rats developed the full spectrum of radium poisoning conditions, including spontaneous fractures and osteogenic sarcomas or bone cancers. However, Evans found that the animals required several hundred times more radium than that which was already known to produce similar conditions in humans.

After several years of effort, Evans finally gave up his experiments because he could not determine from them a permissible dose of radium for man. He concluded, paraphrasing the English poet Alexander Pope (1688–1744), "the proper subject for the study of man is man."[7]

Throughout the 1930s, Evans also worked closely with the famous New Jersey medical examiner and pathologist, Dr. Harrison S. Martland. Evans built and supplied Martland with Geiger counters and other instruments to measure the radium body burdens of the dial painters that he encountered. And Martland referred dial painters to Evans, so that their radium burdens could be measured at MIT.

In late 1940, as the country was gearing up for war, a captain of the U.S. Navy Medical Corps came to see Evans at his laboratory. The captain insisted that Evans quickly provide the Navy with safety standards for the nation's radium dial-painting industry. Further, he threatened that if the young professor did not do so, the captain would see to it that Evans was inducted into the Navy where he would be forced to come up with them.

At the same time, the National Bureau of Standards also began to address the issue of safety standards for the dial-painting industry. The bureau established an advisory committee charged with preparing a handbook on the safe handling of radium compounds. The committee consisted of nine members including Evans, Martland, and Flinn, as well as representatives from the radium industry and government.

On February 26, 1941, the committee met. At the meeting, Evans reviewed the disappointing results of his animal experiments. He then discussed the dial painters he and Dr. Martland had measured—a total of only 27 persons. Evans described the amounts of radium in 20 of the women who were known to be suffering the effects of radium poisoning. He then described the lower body burdens in the remaining seven women who had no apparent injury.

After discussing the cases and noting that the committee was obliged to make a decision, Evans proposed to set a "maximum tolerance level" for radium. Evans said that the level should be set so that the committee (which was all male) would feel perfectly comfortable if their own wives or daughters were subjected to it. Evans then asked each member of the committee if he would be content with establishing the maximum permissible body burden of radium at one ten-millionth of a gram (0.1 microgram). This value was one-tenth the smallest amount of ingested radium known to have adversely affected the health of a dial painter. They agreed unanimously.

Evans then proposed a standard for the permissible level of radon for air concentrations in the radium plants—the first standard for inhaled radioactive gases. It also was unanimously accepted.[8]

On May 2, 1941, the National Bureau of Standards issued the committee's safety recommendations in *Safe Handling of Radioactive Luminous Compound*.[9] Seven months later the Japanese attacked Pearl Harbor, and the United States entered World War II.

With the nation's entry into the war, the military demand for radium dials skyrocketed, and thousands of new dial painters were hired across the country. In 1942, for instance, the U.S. Radium Corporation increased its personnel by 1600%. At its peak wartime production, this company alone employed almost a thousand radium dial painters.[10]

Creation of the radium and radon occupational safety standards would not only protect this new generation of dial painters from undue exposure, but the standards also would prove to be crucial to the Manhattan Engineer District, the United States' secret atomic bomb project.

The Manhattan Project

By the late 1930s, it was clear to many scientists that it was theoretically possible to release atomic energy and produce a tremendously powerful bomb, but no one knew any practical method of doing it. With the outbreak of war in Europe, the United States feared that Nazi Germany might be first to develop such a weapon of mass destruction.

At the time, the threat seemed very real. The Germans had several brilliant scientists working in nuclear physics and chemistry. The two chemists, Otto Hahn (1879–1968) and Fritz Strassmann (1902–1980), for example, had discovered the process of fission in uranium in 1938, and with the occupation of Czechoslovakia in 1939, the Nazis now had a ready supply of uranium from the St. Joachimsthal mines.

To beat the Nazis, the United States began its race to build an atomic bomb. From the beginning, it was clear that developing such a weapon would be an enormous, costly, and complex task—with no guarantee of success. President Harry S. Truman (1884–1972) would later call the massive $2-billion project "the greatest scientific gamble in history."[11]

To succeed, the project would require the expertise of the nation's leading scientists and engineers; tremendous financial and managerial resources; the construction of entirely new cities, industrial plants, and laboratories; the employment of 125,000 workers of all description; and the acquisition and processing of enormous amounts of radioactive material—vastly larger than all the radium ever used.[12]

The project would use two sources of radioactive material to produce the atomic bomb: a rare isotope (or form) of uranium, and a newly discovered man-made element—plutonium.

Initially, the bomb project tried to use uranium-235, the only isotope of uranium that is fissionable (capable of being split and producing an atomic explosion). However, separating and concentrating the rare isotope from the more common form of uranium was very difficult. And after years of effort, only one uranium bomb would ever be produced.

In time, however, the second element would be used. In late 1940, Glenn T. Seaborg (1912–1999), a young chemist at the University of California at Berkeley, identified a trace of a new radioactive element which he later named plutonium (Figure 46). The new element, which could be produced in a nuclear reactor by bombarding common uranium with neutrons, would eventually be found to have several important advantages in making the bomb.

Plutonium was more fissile than uranium-235; therefore, a smaller amount of it was required per bomb, and thus more bombs could be manufactured sooner if plutonium could be produced in quantity. Even more important, the element was less difficult to produce. The new element, however, had one very important disadvantage: nothing was known about it.

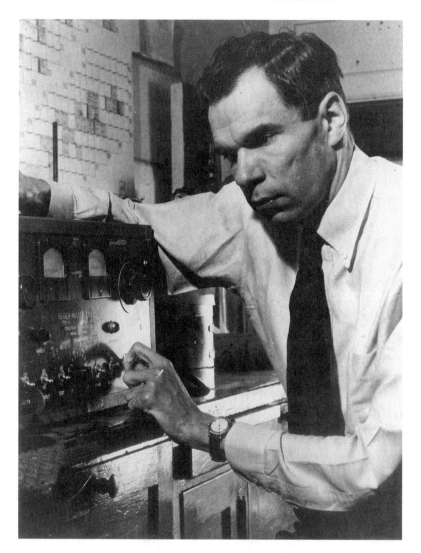

Figure 46. *Glenn T. Seaborg.*

Over the next several years, Seaborg and his assistants produced minute quantities of plutonium. Seaborg soon found that plutonium, like radium, primarily emitted alpha radiation (charged particles composed of two neutrons and two protons)—the most hazardous form of atomic radiation, if such radioactive material enters and remains inside the body.

Although the scientists, physicians, and radiation safety officers of the atomic bomb project expected plutonium to be a major health hazard, they paid very little attention to it before 1944. Because so little of it was produced, other issues seemed much more urgent.[13]

However, just before large-scale production of the element was to begin, Seaborg became alarmed about the potential hazards of the element. He remembered the tragic fate of the radium dial painters, and he was determined not to see the chilling episode repeated. On January 5, 1944, he made the following entry in his diary:

> As I was making the rounds of the laboratory rooms this morning, I was suddenly struck by a disturbing vision. I pictured in my mind the expanded scale of work with solutions containing plutonium that will soon result from the large quantities of plutonium soon to be received from Clinton Laboratories [Oak Ridge, Tennessee]. I visualized beakers of plutonium solutions throughout the laboratory rooms, and it struck me forcibly for the first time that plutonium handling will now no longer be confined to micro quantities manipulated by specially trained experts. Recalling the health problems incurred by workers in the radium dial painting industry, I realized clearly that similar hazards face those of us working with alpha-active material which presents special hazards of ingestion.[14]

Seaborg immediately sent a memo to the chief radiation safety officer about the problem. He explained:

> It has occurred to me that the physiological hazards of working with plutonium and its compounds may be very great. Due to its alpha radiation and long life, it may be that the permanent location in the body of even very small amounts, say one milligram or less, may be very harmful. The ingestion of such extraordinarily small amounts as some few tens of micrograms might be unpleasant, if it locates itself in a permanent position. In the handling of the relatively large amounts soon to begin here [the Metallurgical Laboratory of the University of Chicago] and at Site Y [Los Alamos, New Mexico], there are many conceivable methods by which amounts of this order might be taken in unless the greatest care is exercised.
>
> In addition to helping to set up safety measures in handling so as to prevent the occurrence of such accidents, I would like to suggest that a program to trace the course of plutonium in the body be initiated as soon as possible. In my opinion such a program should have the very highest priority.[15]

To test the element's toxic effects, animal experiments were started. The experiments quickly confirmed that plutonium was indeed biomedically very similar to radium. When ingested, both elements were found to collect in bone. Also, if inhaled, both irradiated the lung where they could cause cancer.

After receiving the results of the experiments, the Manhattan Project established a provisional tolerance level for plutonium based on the standard for radium. Work also began at once to alter the safety equipment and procedures at the project's laboratories and production facilities. Some of the equipment such as glass hoods and ventilation systems would be directly modeled after a new radium dial-painting facility in Boston which Robley Evans helped design.[16]

After a sufficient amount of plutonium was produced and the complex implosion design of the bomb was worked out, the world's first atomic weapon was ready to be test-fired. The test took place on July 16, 1945, in the remote desert at Alamogordo, New Mexico. At 5:30 a.m. the bomb was detonated. The explosion and its resulting blast and shock wave were "unprecedented, magnificent, beautiful, stupendous, and terrifying."[17]

The destructive power of the weapon exceeded all expectations. The 100-foot-high steel tower on which the device was detonated was completely vaporized. The tremendous force of the explosion was so great that it created a crater almost a quarter of a mile in diameter. A steel container weighing more than 200 tons, which was a half-mile from ground zero, was knocked over by the force. And the searing heat from the explosion had turned a large area of desert sand around the crater into a sea of green glass.[18]

Perhaps the most spectacular feature of the bomb, however, was its intense flash, which could be seen 180 miles away. For a few seconds the flash from the explosion was hundreds of times brighter than the midday sun. The light was so brilliant that it temporarily blinded several observers, despite the use of protective black glasses.[19]

Several weeks later, the world's first combat atomic bomb was ready to be deployed. It would be the Manhattan Project's only uranium bomb. The single bomb, carried by a single airplane, would have the destructive force equal to the total bomb load of 2,000 B-29 Superfortresses.[20]

On the morning of August 6, 1945, the bomb was dropped on the Japanese city of Hiroshima. It exploded above the city, and within seconds, five square miles of Hiroshima were completely destroyed. An estimated 70,000 people were instantly killed and another 70,000 seriously injured. Over the years, deaths would continue from the lingering effects of radiation exposure.[21]

After the attack, President Truman informed the American people of the development and immense power of the new atomic bomb. He also grimly warned Japan that if it refused to surrender unconditionally, the United States would attack additional targets with equally devastating results.[22]

At first, the Japanese could not believe that a single bomb could possibly have caused so much destruction. After verifying that indeed it was an atomic bomb, they accused the United States of conducting an atrocity campaign. They called the bomb a "useless cruelty," and they claimed that its use was a violation of international law.[23]

Three days later, on August 9, a second atomic bomb—a plutonium bomb—was dropped on the city of Nagasaki. It totally destroyed three square miles of the city, killing 40,000 people and injuring 60,000 more.[24] Six days later, Japan surrendered and World War II ended. The formal ceremony took place on September 2, 1945.

Shortly after the bombing of Nagasaki, the radiation safety officers at the Hanford Engineer Works wrote to the thousands of employees at the massive plutonium production facility concerning the procedures that had been used to protect their health and safety. Because the project was so secret (even the name plutonium was classified), many of the employees were totally unaware of the nature of the radioactive materials they had been working with. The document stated:

> Previous experience in the radium industry gave us precise knowledge on the limits of protection which would have to be met in dealing with many of the radioactive solids, liquids, and gases developed in our process. Not only have these limits been defined for radium by the National Bureau of Standards, but even the maximum allowable amount of radium which a worker could take into his body without causing injury was known, defined, and could be measured in the body. These limits ... have proved invaluable to us as a guide for our problems at the Hanford Engineer Works.[25]

Without the radium occupational safety standard, which was the direct legacy of the dial painters, tens of thousands of America's atomic bomb workers might have been exposed to dangerously high levels of radioactivity. To the Manhattan Project, the terrible tragedy of the radium dial painters proved to be truly "a most valuable accident."[26] If it had not occurred, the project "might have proved to be a tremendous booby trap."[27]

An official of the U.S. Atomic Energy Commission later stated:

> If it hadn't been for those dial painters, the [atomic bomb] project's management could have reasonably rejected the extreme precautions that were urged on it—the remote-control gadgetry, the dust-dispersal systems, the filtering of exhaust air—and thousands of ... workers might well have been, and might still be, in great danger.[28]

During the Cold War the radium dial workers once again would provide invaluable information, this time on the possible health effects of radioactive fallout.

Chapter 10
Under Radioactive Clouds

A worldwide search is on to find men and women who have lived two or three decades after having received heavy doses of radium radiation. Studies of the victims are expected to take some of the speculation out of present estimates of the danger of radioactivity, including fallout from nuclear weapons tests.[1]

Robert K. Plumb, 1958

Utopian Visions

With the end of World War II, Americans were euphoric. The nation's atomic bomb monopoly would assure a lasting world peace, and atomic energy would create an entirely new world of scientific and technological progress. Numerous scientists, engineers, and journalists speculated that the new abundant, inexpensive, form of energy would revolutionize civilization.

It was predicted that in the near future atomic energy would heat and cool homes, offices, and factories. Atomic energy would enable clean, pollution-free factories to produce enormous quantities of inexpensive goods. Cities would use the energy to melt snow from roadways and heat airport runways to remove fog. And crops would no longer be subjected to the vagaries of the weather; they would be grown indoors under artificial light, so they could be abundant and available all year.

In the more distant future, atomic engines no bigger than a man's fist would be invented, enabling automobiles and other means of transportation to travel for years without refueling. The skies would teem with huge supersonic atomic-powered airplanes, which could cross the Atlantic Ocean in a mere 30 minutes. Atomic rockets would enable man to travel between the planets. Scarce minerals such as gold and platinum would become common. They would be obtained from seawater using cheap atomic power, or they would be created directly by atomic reactors. "Peaceful" atomic bombs would be used as giant earth-moving devices, carving out new channels and harbors, and literally blasting away mountains. A new Panama Canal, for example, could be constructed in weeks instead of years.[2]

To design the new atomic power plants of the future as well as to continue to produce atomic bombs, the wartime Manhattan Project was dissolved in 1946, and the U.S. Atomic Energy Commission (AEC) was established. The new government commission assumed sole ownership of all of the nation's atomic material, as well as the facilities that produced and used it. Under the AEC the former secret atomic research facilities became national laboratories.

The AEC was very interested in studying all aspects of atomic energy, and they sponsored many research projects. The commission was particularly interested in having Robley Evans continue his studies of the radium dial painters. With the increased production of uranium and plutonium, the radium poisoning cases and the crucial 1941 radium occupational safety standard assumed new significance. Workers at the AEC plants were subject to the ingestion and inhalation of small amounts of radioactive dusts and gases, and because the safety standard was based on the experiences of only a small number of dial painters, it had to be reexamined.

To highlight the importance of the dial painters, the AEC published a book of reprints on radium poisoning. The 193-page book consisted of eight articles Dr. Harrison S. Martland had written from 1925 to 1939.[3]

The AEC also provided funds for the Atomic Bomb Casualty Commission (ABCC). Formally established in early 1947, the purpose of the ABCC was to study the long-term medical effects of the atomic bombs dropped on Hiroshima and Nagasaki. It would conduct medical examinations and tests on the bomb survivors and others who lived in the area where the bombs were dropped. The commission would then seek to correlate the effects such as the incidence of leukemia, cataracts, and birth defects with the radiation released from the bombs.[4]

The studies of the radium dial painters and the atomic bomb survivors would complement each other. Studies of the bomb survivors would provide information on the effects of acute, instantaneous exposure to primarily large amounts of gamma radiation, while those of the dial painters would provide information on chronic, long-term exposure to small amounts of alpha radiation. The two long-term studies would form the basis of much of the world's present knowledge of the health risks of atomic radiation.

The Cold War and Radioactive Fallout

By the late 1940s, America's postwar euphoria was fading as the Cold War was beginning. The Soviet Union and its allies were emerging as a threat to the United States and the rest of the free world. In 1948, Czechoslovakia fell to a Soviet-inspired takeover, and communist East Germany blockaded West Berlin. In 1949, the world was shocked when the Soviet Union detonated its first atomic bomb. That same year, communist forces took over China. In 1950, communist North Korea invaded South Korea. These events would trigger an unprecedented arms race between the United States and the Soviet Union.

Figure 47. *The first atomic test in Nevada.*

To surge ahead of the Soviets, President Truman in early 1950 ordered the AEC to develop a superbomb or hydrogen (thermonuclear) bomb. This proposed bomb would be a thousand times more powerful than the one dropped on Hiroshima. It was hoped that this new weapon would protect the free world from further communist aggression.[5]

To speed up weapons development and testing, the AEC established a new atomic proving grounds in Nevada. Previously, all military and AEC tests had taken place in the South Pacific. The vast Nevada Test Site, located 70 miles northwest of Las Vegas, would provide greater security and be much more accessible.[6]

In January 1951, the first atomic test was conducted at the site (Figure 47). The early morning blast shook buildings, knocked people out of bed, and shattered windows in downtown Las Vegas (Figure 48). The blast was seen as far away as mid-Arizona, south-

***Figure 48.** A 1950s postcard advertising the "Up and Atom" City of Las Vegas.*

western Utah, and southeastern California.[7] A few days later, increased radioactivity levels were detected in Chicago.[8] Radioactive snow fell across the eastern United States and Canada.[9] And radioactive dust spoiled sensitive photographic paper and emulsions at the giant Eastman Kodak plant in Rochester, New York.[10]

Over the next 12 years, the AEC would detonate a total of 100 aboveground explosions at the Nevada site. All sorts of nuclear warheads, triggering devices, and delivery systems would be tested. Tactical nuclear weapons also would be tested, including atomic artillery shells, mines, and small bombs for airplanes, rockets, and guided missiles. Tens of thousands of troops would take part in the tests, some advancing to within several hundred yards of ground zero shortly after the explosions.

In November 1952, the AEC successfully tested a prototype hydrogen bomb in the South Pacific. America's superbomb monopoly, however, was very short lived. Ten months later, the Soviet Union detonated its first hydrogen bomb in Central Asia.

With the Soviets again breaking the U.S.' monopoly, the arms race greatly accelerated. The United States was very concerned that if the Soviet Union surged ahead in its design and production of weapons, it might be tempted to launch a nuclear war.

During the mid-1950s and continuing until 1963, when the United States and the Soviet Union, as well as France and Great Britain, signed the Limited Test Ban Treaty, hundreds of aboveground tests were conducted.[11]

Each test sent tons of radioactive debris and bomb by-products boiling thousands of feet into the sky and drifting across the country and world. The material would gradually come back to earth as radioactive fallout. The fallout consisted of many radioactive elements and isotopes, some of which did not occur in nature, and whose long-term health effects were totally unknown.

One of the most hazardous of these elements was the new radioisotope strontium-90. Unlike many other radionuclides created by the explosions, strontium-90 persisted for decades (it has a radioactive half-life of 30 years). It was chemically similar to calcium, and, like radium, when it was ingested, it tended to concentrate in bone. Furthermore, it was particularly dangerous to the developing bones of young children.

With each test, strontium-90 was increasing in the world's food chain and moving through it to man. As the radioisotope fell back to earth, it would be deposited in soil where it would be absorbed by plants and animals. Cows eating grasses containing the radioisotope would concentrate it in their milk. And humans eating contaminated cereals, vegetables, and dairy products would ingest it, and it would be deposited in their bones.[12]

The AEC was very concerned about the health risks of strontium-90. In 1953, for example, it established a global network (Project Sunshine) to secretly collect tissue and bone samples from cadavers, especially from children, in order to measure the amount of the radioisotope being absorbed by humans. Over several years, more than 1,500 samples would be gathered and analyzed.[13]

At first, few people outside of the AEC were concerned about the health risks from fallout. However, in 1954, a controversial debate would begin. In March, when the United States was conducting a hydrogen bomb test in the South Pacific, over 300 people were exposed to high levels of fallout. Among them were 23 Japanese fishermen aboard a tuna trawler which had inadvertently entered the test site. The fishermen, who were 80 miles from the blast, suffered burns and radiation sickness from fallout. Several months later, one of them would die. The Japanese and much of the world were shocked and outraged by the incident.[14]

The Chairman of the AEC, Lewis Strauss (1896–1974), a self-made Wall Street millionaire, who was a rear admiral in the naval reserve and an ardent anticommunist, would later issue a statement saying that all of the injured individuals were recovering. He also would assure the public that fallout from the Nevada tests was "far below the levels which could be harmful in any way to human beings, animals, or crops."[15]

Over time, the controversy over the risks of fallout would intensify. Several prominent scientists, the most outspoken being the Nobel laureate chemist Linus Pauling (1901–1994), would make dire predictions of its long-term effects. Pauling would repeatedly warn that fallout would cause large numbers of radiation-induced leukemias, bone cancers, and other diseases, as well as serious mental deficiency and physical defects in future generations of children.[16]

In sharp contrast, the AEC claimed the health risks from fallout were insignificant. Willard Libby (1908–1980), a chemist and AEC commissioner who would later win a Nobel Prize for his work in radiocarbon dating, repeatedly stated that the radioactivity from fallout was less than that from wearing a radium dial watch. He also argued that the risks were very small compared to the "terrible future we might face if we fell behind in our nuclear defense effort."[17]

In the fall of 1956, the controversy would become an important political issue. In a nationally televised speech, the Democratic Presidential Candidate Adlai E. Stevenson (1900–1965) called for an end to hydrogen bomb tests. In his speech, Stevenson described the perils of fallout, saying:

> ... strontium-90 ... is the most dreadful poison in the world. For only one tablespoon equally shared by all the members of the human race could produce a dangerous level of radioactivity in the bones of every individual. In sufficient concentrations it can cause bone cancer and dangerously affect the reproductive processes.[18]

The Republican incumbent, President Dwight D. Eisenhower (1890–1969), however, opposed a unilateral end to testing. Despite the growing public apprehension over fallout, the public reelected Eisenhower in a landslide. He even won a majority of votes in Nevada and Utah, the two states receiving the greatest amount of fallout.

That same year, the AEC formed a committee to consider the possible long-term health risks of continuous exposure to strontium-90. Because of its similarity to radium, the committee recommended that the dial painter cases be restudied.[19]

The radium dial painters, who were the largest known group in the world with the oldest internal exposure to radioactivity and the longest retention of it, would provide a unique and invaluable source of information on the likely long-term health effects of chronic exposure to strontium-90. The study of the dial painters' experience would provide vital insight with "implications for hundreds of millions of people all over the world."[20]

One AEC official described the importance of the study, stating:

> Something that happened far in the past is going to give us a look far into the future. Why, when these people took in their radium, there was no such thing as strontium-90, and yet they may help us determine today how much of it children can safely consume. The way I see it, we're trying to follow up a wholly unintentional experiment that has taken on incalculable value.[21]

To find as many of the former dial painters as possible, in 1957 the AEC sponsored an exhaustive search. It contracted with MIT, the New Jersey State Department of Health, and Argonne National Laboratory.

A group at MIT, under the direction of Robley Evans, would study the dial painters residing in New England; the New Jersey health department would study those within their state and in metropolitan New York City; and Argonne National Laboratory, which is located near Chicago, would study those in the Midwest. In addition, all three would attempt to find and study people given radium therapeutically, as well as chemists and laboratory technicians who worked with the element.[22]

Specifically, MIT would attempt to identify former dial painters from the Waterbury Clock Company; the New Jersey health department, from the U.S. Radium Corporation; and Argonne National Laboratory, from the Radium Dial Company.

After identifying the painters and others exposed to radium, the three institutions would attempt to locate, interview, conduct medical examinations and tests, measure their radium body burden, and monitor them over time to identify any changes in their health status.

In November 1957, the New Jersey project started as a feasibility study. In March 1958, the ongoing study began. To conduct the study, the state health department formed the New Jersey Radium Research Project, which set up its headquarters in West Orange, New Jersey.

Locating the former dial painters would prove to be a daunting task. More than a quarter century had passed since the women were employed as dial painters. The U.S. Radium Corporation had few employment records: Social Security did not exist when the women worked. Many of them had changed their names through marriage. And some of them had moved away, or had died elsewhere.

To track them down, the New Jersey project's staff issued press releases and worked closely with the news media. They contacted local medical societies and published articles in the New Jersey medical society's journal asking physicians to review their records and to refer patients to them. The staff also checked old newspapers, city directories, employment lists, telephone books, and marriage and death certificates. In addition, the project used the services of a state police special investigator. The detective, who was assigned to the project, would prove invaluable in searching motor vehicle and court records for clues as to their whereabouts.

From 1957 until around 1960, the New Jersey project concentrated on identifying and locating the former workers. From 1959 through 1965, it concentrated on collecting medical, dental, and laboratory data from those individuals willing to participate in the study.

After a decade of intense effort, the New Jersey project ended in 1967. In total, it identified almost a thousand persons who were exposed to radium, 520 of whom were dial painters. All of the project's files would eventually be transferred to Argonne National Laboratory.[23]

Before the AEC-sponsored study of the dial painters, Argonne National Laboratory had conducted a number of studies of radium. As early as 1946, researchers at the laboratory had attempted to remove radium from the body of a man who had worked with the element for many decades. A few years later, it had conducted a medical follow-up study of former psychiatric patients from the Elgin (Illinois) State Hospital who had been given radium injections in the 1930s. It also had studied the amount of radium in municipal drinking water. As part of the study, Argonne researchers had even tested inmates at the nearby Stateville Prison in Joliet, Illinois, whose well water had high concentrations of the element. The laboratory had already identified a number of former dial painters from the Radium Dial Company for study.[24]

Argonne was also uniquely equipped to measure the dial painters. In 1954, it had assembled the nation's first whole-body counter or "iron room" to measure very small quantities of radioactive materials in the body.[25] In addition, the laboratory had constructed an extremely sensitive radon breath analyzer that could detect as little as one ten-billionth of a gram of deposited radium.[26]

Like the New Jersey project, researchers at Argonne were also faced with difficulties in identifying the former painters. The Radium Dial Company had moved several times, and few employee lists existed. The researchers, however, were greatly helped by the discovery of several group photographs of the company's employees. One photo in particular showed all the dial painters at their desks in a large room.

Argonne had the photographs reproduced and enlarged. And they incorporated them into their interviews with the former dial painters. The photos would prove to be important memory aids. Many of the women identified coworkers, some of whom they had not seen in decades. Eventually, researchers at Argonne would identify, and locate, several hundred dial painters from the Radium Dial Company, and they would measure the radium content of many of them.

During the study, MIT also was very successful in locating former dial painters. Researchers discovered an old employment list from the Waterbury Clock Company, which proved very useful. They also found additional dial painters and radium workers previously employed by the New England Watch Company (also located in Waterbury, Conn.), and the Standard Chemical Company in Pittsburgh. In total, the MIT group would identify over a thousand people exposed to radium. Of these, over 600 were dial painters.[27]

A National Center

In 1967, Robley Evans, after spending 35 years at MIT, was planning to retire. However, he was very concerned that with his retirement the studies of the radium dial painters would end. He believed that in order for them to successfully continue, they should be consolidated under the direction of a single national research organization with assured long-term government funding. Evans proposed that the new organization be called the National Center of Human Radiobiology.

To establish such a center, Evans submitted a proposal to the AEC's Advisory Committee for Biology and Medicine. In it, he forcefully argued that in the future more radioactive material would be used by nuclear power plants, industry, and in weapons production, and therefore it was prudent, and indeed a moral obligation to future generations, to learn as much as possible about the effects of radiation. He called the radium dial painters a unique and invaluable group, whose experience would never be duplicated. He stated that the studies of them were just beginning to show the late effects of their exposure. Evans argued that the only way to truly know if all of the effects had been observed was to continue the studies of them for their full life spans, and that detailed laboratory studies should be performed on their tissues after their deaths.[28]

The AEC approved Evans' proposal in 1969, and the Center for Human Radiobiology (CHR) was officially established at Argonne National Laboratory. The new national center was designated by the AEC to conduct long-term follow-up studies of not only the dial painters, but all individuals with internally deposited radioactivity.

By 1970, all of the dial painter records gathered by MIT and the New Jersey project were transferred and consolidated at the CHR. In 1971, the center began actively recruiting dial painters to come to Argonne to be examined. While they were at the laboratory, they would receive a comprehensive physical examination, a radon breath test, a whole-body count, a measurement of bone mineral mass, and an examination of their chromosomes.[29]

Over the next several years the CHR greatly expanded, and hundreds of new dial painters were identified and measured. The center also began studying 5,000 former thorium workers who produced gas mantles at the Lindsay Chemical Company in Chicago, and later at West Chicago. In addition, it identified and conducted investigations of individuals exposed to plutonium, americium, and other radioactive elements.

In late 1975, the CHR moved into a new, specially constructed $1.3 million facility at Argonne. The facility contained offices for the staff, specially designed rooms for record storage, and a morgue.

To make as accurate measurements as possible of radioactivity in the living, the CHR specially designed several whole-body counters or "iron rooms." The counters were six-feet-wide, seven-feet-long, and eight-foot-high cubicles constructed of eight-inch-thick steel. To reduce background radioactivity as much as possible, the insides of the cubicles were lined with lead and they were supplied with air from which the radon had been removed. Finally, to block out cosmic radiation from outer space, the counters were housed in a specially built concrete vault with three-foot-thick walls, and the vault was covered with 11 feet of earth.

The person to be measured would first change clothes outside the vault and don a gown and paper shoes so they would not carry radioactive dust into the counter. They would walk into the vault and enter the counter through a large sliding steel door, which weighed 9,000 pounds. The individual would then sit on a specially designed reclining steel chair. The door would slowly close, and the individual would remain seated inside the counter for approximately 20 to 30 minutes. The amount of gamma radiation emitted from the person's body would then be carefully calculated, and the quantity and type of radioactive materials determined.[30]

To make as accurate measurements as possible of radioactivity in the dead, the CHR had a fully equipped morgue constructed for its use. It would exhume over 100 individuals, and their remains would be examined there. Also, about two dozen willed bodies would be autopsied in the morgue, providing the center a unique opportunity for detailed pathological and radiological studies.[31]

The CHR would use the valuable data it gathered to write hundreds of scientific and medical articles. The articles would report changes in the dial painters' bones, blood, and immune systems. Others would address the specific incidence of such diseases as bone cancer, head and neck cancer, breast cancer, multiple myeloma, and leukemia among the workers.[32]

In addition, other articles were written on the health effects of thorium, plutonium, and other radionuclides. Several articles discussed the effects of fallout from the Chernobyl reactor accident, while others examined the recently discovered environmental health problem of indoor radon.[33]

Beginning in 1983, budgetary restrictions began to slow the efforts of the center, and fewer and fewer patients were brought to the laboratory. Despite decreased funding, however, the center's staff continued to record the deaths of cases, and to write and publish articles.

In 1990, the Center for Human Radiobiology was downgraded to the status of a program. A few years later, the program was terminated. It officially closed on September 30, 1993.[34]

By the end of the program, a staggering total of 5,650 radium dial painters were identified and 1,934 of them located and measured for radium body content; 485 chemists and laboratory technicians who worked with radium were identified and 282 located and measured; and 357 persons given radium therapeutically were identified and 282 measured.[35]

Chapter 11

Conclusion

Just as shadow is the result of illumination, disasters are the inevitable sequel of technological innovation.[1]
                                    Wade Roush, 1993

Since the first tragic deaths of the radium dial workers more than three-quarters of a century ago, there has been an enormous number of changes. Radium is no longer used on watch dials, in consumer products, or by medicine. It has been replaced either by cheaper, less toxic man-made radioactive materials, or its use has been totally eliminated. Today, the once precious and expensive element is now little more than worthless hazardous waste, which is very difficult and costly to dispose of.

More comprehensive radiation occupational protection standards have evolved, going from being exclusively concerned with immediate and acute health effects, to long-term and genetic effects, and finally to a risk-based approach taking into consideration additional factors such as loss of life expectancy and morbidity. Current protection standards not only address the established limits of the doses to individuals, but also such factors as the net benefit of working with radioactivity. Now the standards recommend that all exposures must be kept as low as reasonably achievable, taking economic and social factors into account.[2]

Laws and regulations protecting workers also have greatly increased. During the time of the dial painters, each state enacted its own workplace health and safety laws, which were enforced piecemeal, if at all. Today, the federal Occupational Safety and Health Administration (OSHA) regulates and enforces national workplace safety and health standards.

In the past, workers were rarely told about the harmful materials to which they were exposed. Under OSHA's regulations, employers are now required to inform workers of the presence of all hazardous chemicals and toxic substances. Employers also must label all hazardous substances; train workers in the proper means of handling them; develop a written hazardous communication program; and have a material safety data sheet for each hazardous substance they use.

Unlike the dial painters, workers now have the legal right to see all records of their exposure to harmful materials, including the results of all medical tests conducted by their employers. Employers must also retain all workers' medical records for at least 30 years after termination of their employment.[3]

Despite these changes, workplace hazards still continue to inflict a tremendous toll. Tens of millions of workers are exposed daily to unsafe work situations, harmful chemicals, and toxic substances. In one year alone (1994), the nation's employers reported 6.3 million work injuries and 515,000 cases of occupational illnesses.[4] According to the National Institute for Occupational Safety and Health, each day in the nation, on average, 137 persons die from work-related diseases.[5] Of these, 55 die from cancers caused by exposure in the workplace.[6]

Still, many things remain the same. Over the years, state workers' compensation programs have changed very little. Many of the state programs continue to have short statutes of limitations, which preclude diseases discovered long after workers have left their jobs. Further, the programs suffer from lengthy delays and long negotiations, and all of them pay only a small, inadequate amount of benefits to disease victims.

Like the radium dial painters, in order to receive compensation, workers still have the difficult burden of proving their work exposure caused the specific disease from which they are suffering. Workers must show that their exposure probably caused their illness; it is not enough to show that it possibly caused it. Many workers, however, do not know what hazardous substances they are exposed to, nor do they recognize the illness they are suffering from is work-related. Even when they are aware of this, many believe they cannot prove it. In addition, workers are often reluctant to file compensation claims against current employers, fearing it may jeopardize their jobs.[7]

Although the widespread use of radium medicines and tonics ended with the death of Eben Byers, a new radium fad would arise in World War II and continue until the mid-1960s. During the war, many U.S. pilots, air crew members, and submariners experienced inner ear problems because of rapid changes in air pressure. To relieve pain and to return the servicemen to active duty as quickly as possible, many were treated with radium. The treatment consisted of inserting a radium nasal applicator (a slender, 6 3/4-inch long metal rod with a small, sealed radium capsule at its tip) into each nostril and extending them through the floor of the nose to the back of the mouth. The devices were then left there for 6 to 12 minutes to literally burn away tissue around the eustachian tubes. The treatment delivered very high doses to tissues immediately adjacent to the devices, and smaller doses to other tissues and organs farther away.

Starting in the late 1940s, radium nasal treatment became very popular among physicians and the general public. Many mistakenly believed the supposedly harmless treatment corrected middle-ear defects in children and therefore prevented deafness in later life. Eventually, the treatment became a routine medical procedure. It would not be until the mid-1950s, with growing evidence of the long-term harmful effects of radiation exposure, especially among children, and the rise of alternative treatments (i.e., antibiotics, and use of tympanotomy tubes) that the treatment slowly began to fall out of favor.

While the exact numbers are unknown, an estimated 15,000 to 20,000 servicemen, and 500,000 to 2.5 million children received radium nasal treatments. These individuals are now at increased risk of developing various thyroid diseases, tumors of the head and neck, and brain cancer.[8]

Quack radioactive treatments continue to be used to this day. Every year, over 5,000 people flock to a half-dozen former uranium mines in western Montana, which have been converted into "health spas." They sit for hours in the mines to breathe high concentrations of radon, they cover themselves with radioactive mud, and they drink radioactive water in hopes of being cured. Like the charlatans of the past, the spas falsely claim that these potentially dangerous treatments will cure everything from diabetes to rheumatism.[9]

Although the nation's once prominent radium industry has passed into history, it has left behind a legacy of radioactively contaminated buildings and communities. From the time of radium's discovery, very little attention was paid to the possibility of contamination. The element was considered too precious for any company or individual to waste or carelessly discard, and for many decades there were no regulations governing radium's use or disposal.

Beginning in the 1950s, however, several radium-contaminated buildings were "discovered." Initially, they were viewed as rare atomic novelties. During the 1970s, many more radium-contaminated sites were found, and several existing dial-painting plants were forced to close because of contamination. In the 1980s, many of the sites and plants were included on the U.S. EPA's Superfund list for remediation. During the 1990s, the cleanup of many of the sites continues.

One of the most controversial and costly sites was the former U.S. Radium Corporation plant of Orange, New Jersey. When the plant processed carnotite ore to produce radium from 1917–1926, it discarded large amounts of radioactive waste or "mill tailings" on the plant's property. As more of the tailings were produced, however, they were used as fill in low-lying areas in the surrounding communities of West Orange, Glen Ridge, and Montclair. Later, houses were built atop or around the fill.

Eventually, more than 200 acres of radioactively contaminated soil were found in the four communities. In some cases, the tainted soil extended to a depth of 15 feet or more. In total, almost 750 homes were determined to have unsafe levels of gamma radiation and radon due to the contaminated soil.

In 1985, the New Jersey Department of Environmental Protection (DEP) began a pilot project to remove the contaminated soil from around 12 homes. The project quickly turned into a nightmare. After removing and filling thousands of drums with the tainted soil, the DEP was faced with the very difficult task of finding a permanent disposal site for it. No community in the state would accept it. The DEP tried to ship the soil out-of-state. However, they again failed. After drawing up an agreement with a Nevada dump site, the governor and state legislator of Nevada passed a law blocking the transport of the soil into that state. Eventually, after much effort and frustration, the soil was sent to Oak Ridge, Tennessee, where it was mixed with highly radioactive wastes from nuclear power plants. The mix then was considered high-level nuclear waste and therefore qualified for disposal at the U.S. Department of Energy's Hanford, Washington, facility, where it was eventually buried.[10]

In 1990, the U.S. EPA finally agreed to remove all of the contaminated soil and restore the properties within the four communities. All of the contaminated soil would be shipped to a site in Utah for final disposal. The federal cleanup was estimated to cost from $250 to $350 million, and take 10 years to complete.[11]

By 1996, 300 houses were fully cleaned, and 80,000 cubic yards of contaminated soil— 5,000 large truckloads—were shipped out-of-state. Today, the complicated and enormously expensive cleanup continues.[12]

Even though many contaminated sites have been found, because radium was so widely used in the past, there undoubtedly are many factories, chemical processing plants, laboratories, physician offices, watch repair shops, and homes contaminated with the element which have not yet been identified, which may pose serious public health problems. The former Waterbury Clock Company of Waterbury, Connecticut, for example, was only recently (1998) found to be radioactively contaminated.[13]

In a broad context, the radium dial tragedy can be viewed as a public health case study of the unanticipated consequences of scientific discovery and the application of new scientific knowledge in the areas of technology, industry, and health care. The tragedy provides several insights and recommendations for the future.

Although modern society tends to define progress in terms of scientific discoveries and the resulting technologies and products they create, society must also recognize that progress has a high price. It creates new complex public health problems with harmful, and even deadly consequences. Indeed, the problems caused by progress such as air and water pollution, toxic chemical waste, and radioactive contamination, have emerged as a major source of environmental and health hazards.[14]

When a technological disaster occurs, it is imperative that corporate managers quickly accept responsibility and admit the mistake. After learning of the disaster, managers must immediately stop using the harmful technology, end the production of the hazardous product, issue appropriate warnings, and initiate timely recalls. In addition, managers must thoroughly learn from the disaster, and they must ensure that it never happens again.

Victims of technological disasters must not be blamed for them. Companies sometimes attempt to justify disasters by finding defects in the victims. The radium companies, for example, claimed that the dial painters became sick because they were cripples who could not find work elsewhere. Later, the women were accused of developing radium poisoning because they were greedy, painting more dials and thus pointing their brushes more frequently than other dial painters who did not become sick.

Existing laws must be reformed and new legislation promulgated to provide more just, and faster compensation to victims of disasters. The victims should not have to spend years proving the cause of their injuries. To start the process, antiquated state workers' compensation programs must be updated, broadened, and made more consistent across states.

All members of society, including workers, as well as corporations and public health departments must continuously and vigilantly monitor all new technologies and products for possible adverse effects. Society also must support the development of research and laboratory methods to predict emerging risks. These methods must identify the possible harmful synergistic effects of combining new technologies with existing ones, and they also must determine how new products may be inappropriately used, and even purposefully abused.

As a society and as individuals, we must give greater consideration to the long-term impacts of our actions on future generations. In retrospect, the use of radium in consumer products serves as an exemplar of shortsightedness. It was foolish to use radium, an element with a radioactive half-life of 1,600 years, on the dials of watches, which only lasted a year or two. Now, a new generation, which never used radium products, must face the health hazards and pay the enormous costs of cleaning up the wastes of the past.

Today, few reminders of the tragic story of the radium dial workers remain. No monuments have ever been erected to honor their memory. Even in the cities where the tragedies took place, local histories fail to mention them. Despite this sad neglect, the suffering and deaths of these workers greatly increased the world's knowledge of the hazards of radioactivity, ultimately saving countless lives of future generations. This clearly is their most enduring and greatest tribute.

Appendix

Table A1. Radium-Caused Mortality Among Dial Painters and Other Workers of Essex County, New Jersey

Subject*	Period of Exposure	Age at Death	Year of Death	Listed Cause of Death
1	1917–1922	25	1922	Ulcerative stomatitis, Syphilis
2	1917–1922	24	1922	Pernicious anemia
3	1918–1922	22	1923	Primary anemia, Vincent's angina, Pneumonia
4	1918–1921	21	1923	Phosphorus poisoning, Complete necrosis of the lower jaw
5	1918–1922	20	1924	Sarcoma of right knee
6	1917–1920	25	1924	Necrosis of jaw and maxilla, Anemia
7	1913–1917, 1920–1924	36	1925	Lobar pneumonia, Acute myocarditis
8	1912–1920, 1920–1925	36	1925	Pernicious anemia
9	1917–1925	35	1925	Pernicious anemia, Necrosis of jaw, Sepsis
10	1919–1923	24	1925	Aplastic anemia, Jaw necrosis, Terminal pneumonia
11	1917–1920	24	1927	Chronic nephritis, Chronic myocarditis, Hypostatic pneumonia
12	1917–1919	33	1927	Osteogenic sarcoma of scapula
13	1914–1927	44	1927	Chronic progressive anemia, Hypostatic pneumonia
14	1913–1921	45	1928	Aplastic anemia, Terminal sepsis, Hematosis and Bronchopneumonia
15	1917–1919	29	1929	Osteogenic sarcoma of pelvis
16	1918–1922	27	1930	Sarcoma of pelvis with metastases to lungs and ethmoid
17	1917–1921	31	1931	Osteogenic sarcoma of pelvis
18	1919–1923	35	1932	Ulcerative and necrotizing colitis
19	1917–1921	30	1933	Sarcoma of left femur
20	1917–1920	34	1933	Periosteal sarcoma of right femur
21	1917–1921	36	1935	Bilateral mastoiditis, Sinus thrombosis, Venous sepsis
22	1919–1921	31	1935	Lobar pneumonia, Spontaneous pneumothorax due to metastases of osteosarcoma of arm

Appendix

23	1917–1924	40	1936	Sarcoma of pelvis, Osteogenic osteolytic
24	1919–1924	38	1938	Sarcoma of left orbit
25	1917–1922	37	1939	Sarcoma of femur
26	?–1921	37	1940	Osteogenic sarcoma of left talus
27	1917–1924	41	1941	Sarcoma of right hip and pelvis
28	?	52	1941	Primary carcinoma of bronchus
29	1918–1921	39	1942	Sarcoma of femur
30	1919–1924	49	1943	Epidermoid carcinoma of mastoid
31	1916–1922	52	1945	Carcinoma of nose, Epidermoid carcinoma of ethmoid sinus
32	1918	47	1945	Osteogenic sarcoma of leg
33	1917–1919	51	1946	Sarcoma of femur
34	1916–1919	48	1949	Necrosis of maxillae and temporal bone, Carcinoma, Bronchopneumonia
35	1917–1919	51	1951	Epidermoid carcinoma, Radiation osteitis
36	1916–1922	55	1956	Osteogenic sarcoma
37	1917–1918	56	1957	Adenocarcinoma of sphenoid sinus invading optic nerve and orbit
38	1916–1918	57	1958	Osteogenic sarcoma of femur
39	1917	58	1958	Adenocarcinoma of sphenoid sinus
40	1919–1920	70	1959	Adenocarcinoma of paranasal sinus, Terminal bronchopneumonia
41	1918	61	1962	Metastatic osteogenic sarcoma, Osteogenic sarcoma
42	1917–1919	73	1965	Metastatic carcinoma, Cancer of transverse colon
43	1920–1927	67	1966	Carcinoma of lung with metastases
44	1917	66	1968	Metastatic cancer from anaplastic bone sarcoma of sternoclavicular joint
45	1916–1918	69	1968	Osteogenic fibrosarcoma and pleural effusion
46	1918–1919	75	1978	Fibrosarcoma of right femur
47	1918–1919	80	1983	Fibrosarcoma of knee

*All subjects worked only at the U.S. Radium Corp. of Orange, NJ, except subject 8, who first worked for the Carnotite Reduction Co., IL, from 1912–1920, and subjects 19, 25, and 27, who worked for Luminite Corp., NJ, after leaving U.S. Radium Corp. All subjects were dial painters, except 2, 8, and 43, who were chemists; 14 and 28 were managers; and 13 was a clerk.

Table A2. Radium-Caused Mortality Among Dial Painters of New Haven County, Connecticut

Subject*	Period of Exposure	Age at Death	Year of Death	Listed Cause of Death
1	1921–1925	21	1925	Secondary anemia, Ulcerative stomatitis, Osteomyelitis of upper alveolar
2	1923–1927	25	1927	Chronic poisoning by radioactive paints
3	1922–1925	23	1927	Radium poisoning, Osteomyelitis of mandible and maxilla
4	1923–1924	21	1929	Uremia septicemia, Radium poisoning
5	1924–1927	19	1930	Radium poisoning, Brain abscess
6	1923, 1926	20	1930	Osteogenic sarcoma of femur
7	1922–1927	23	1931	Osteogenic sarcoma of ribs
8	1923–1924	23	1931	Spindle cell sarcoma of the shoulder
9	1924–1925	22	1931	Bone necrosis (maxilla)
10	1923–1926	25	1931	Anemia, Secondary to osteosarcoma
11	1923–1925	27	1934	Osteosarcoma (iliac bones)
12	1923–1928	26	1935	Necrosis of mandible, Osteogenic sarcoma (humerus)
13	1919–?	33	1935	Osteomyelitis of superior maxilla and mastoid
14	1921–1934	32	1936	Radiumgenic sarcomatoid tumor of jaw
15	1920	33	1937	Osteomyelitis of superior and inferior maxillae, Anemia
16	1924–1928	27	1937	Osteomyelitis of cranium, femur, and mandible
17	1924–1928	32	1940	Osteogenic sarcoma
18	1920–1924	42	1943	Radiogenic osteosarcoma of femur
19	1920–1933	40	1943	Osteosarcoma of leg
20	1924–1930	35	1944	Radium poisoning
21	1923–1925	35	1944	Necrosis of temporal bone, Brain abscess
22	1924	39	1945	Sarcoma of tibia
23	1921	47	1945	Intracranial pressure, Osteogenic sarcoma

Appendix

24	1922–1930	42	1946	Osteogenic sarcoma of rib
25	1924–?	39	1947	Osteogenic sarcoma of pelvis, hip, and mandible
26	1924–1940	46	1955	Osteogenic sarcoma of humerus
27	1919–1920	78	1956	Sarcoma of leg with metastasis to liver
28	1920–?	61	1967	Sarcoma of hip
29	1921–1922, 1926–1927	68	1969	Osteosarcoma of femur
30	1920–1928	73	1978	Epidermoid carcinoma, metastatic

All subjects worked only at the Waterbury Clock Co. of Waterbury, CT, except subject 7, who first worked at the Ansonia Clock Co., NY, from 1922–1924, and subject 14, who worked at the New England Watch Co., CT, for one year—1921.

Table A3. Radium-Caused Mortality Among Dial Painters of LaSalle County, Illinois

Subject*	Period of Exposure	Age at Death	Year of Death	Listed Cause of Death
1	1923–1927	24	1927	Streptococcal septicemia, Infection of face
2	1923–1929	23	1929	Diphtheria, Bronchopneumonia
3	1923–1928	21	1930	Sarcoma of retroperitoneal lymph glands
4	?–1924	26	1930	Sarcoma of hip
5	1920–1921	30	1931	Radium poisoning, Sarcoma of pelvic
6	1923–1925	27	1934	Sarcoma of femur
7	1923–1929	29	1936	Spindle-cell sarcoma of neck
8	1922–1931	35	1938	Osteogenic sarcoma, Aplastic anemia, Necrosis of jaw
9	1922–1936	32	1939	Chronic granulocytic leukemia
10	1919	36	1942	Sarcoma of fascia of thigh
11	1924–1936	36	1945	Osteogenic sarcoma of femur
12	1924–?	39	1946	Sarcoma of humerous
13	1925–1928	38	1946	Tumor base of brain petrous portion of temporal bone
14	1925–?	45	1946	Aplastic anemia
15	1921–1927	48	1951	Radiation osteitis
16	1922	46	1953	Fibrosarcoma of femur
17	1922–1928	48	1953	Carcinoma of ear and mastoid
18	1924–1940	47	1954	Fibrosarcoma of ischium
19	1925–1926	51	1955	Carcinoma of mastoid
20	1922–1926	59	1956	Sarcoma of hip
21	1923–1925	53	1956	Fibrosarcoma of mastoids
22	1922–?	54	1959	Osteogenic sarcoma
23	1924–?	63	1960	Rhabdomyosarcoma of maxilla
24	1923–1946	55	1961	Epidermoid carcinoma of mastoid

25	1922–1923, 1935	59	Fibrosarcoma of femur
26	1923–1925	62	Pleomorphic sarcoma
27	1924–1925	56	Squamous cell carcinoma of gingiva
28	1923–1935	64	Squamous cell carcinoma of mastoid
29	1922–1923	68	Carcinoma of mastoid
30	1923–1927	67	Generalized squamous cell carcinoma
31	1924–1925	79	Epidermoid carcinoma of mastoid
32	1924–1925	73	Carcinoma of sphenoid sinus
33	1923–1926	77	Carcinoma of mastoid
34	1924–1926	80	Cancer of auditory canal
35	1919, 1923–1932	83	Fibrosarcoma of ankle

All subjects worked only at the Radium Dial Co., of Ottawa, IL, except 9, 18, 24, 25, and 28, who worked for both Radium Dial and Luminous Processes, Inc., of Ottawa and New York City.

Notes

Introduction

1. Bale T. A brush with justice: the New Jersey radium dial painters in the courts. *Health/PAC Bull.* 1987;17(5):21.
2. Sharpe WD. The New Jersey radium dial workers: seventy-five years later. In: Sheehan HE, Wedeen RP, eds. *Toxic Circles: Environmental Hazards From the Workplace Into the Community.* New Brunswick, NJ: Rutgers University Press; 1993:138–167.
3. Miles AEW. Phosphorus necrosis of the jaw: 'phossy jaw.' *Br Dental J.* 1972;133:203–206.
4. See for example the various correspondence between Wiley KGT and Roeder A, January 9, 17, 19, and March 3, 1925. MSS 18509, Reel No. 86. Located at: National Consumers' League Papers, U.S. Library of Congress, Washington, DC.
5. New Jersey radium cases—the lesson. *Am Labor Leg Rev.* 1928;18:388.
6. Poison paintbrush. *Time.* June 4, 1928:40.
7. Mme. Curie holds out scant hope for girls facing radium death. *New York Evening Journal.* May 26, 1928:A1:1.
8. Goldmark J. *Impatient Crusader: Florence Kelley's Life Story.* Urbana, Ill: University of Illinois Press; 1953:197.
9. Radium handlers sought in continuing study. *Medical Tribune.* October 22, 1962:22.
10. A list of all known deaths is contained in the Appendix Tables A1–A3.
11. See for example Griffin F. Society of the living dead. *Star Weekly* [Toronto, Can.]. April 23, 1938:3; Main J. Fifteen walking ghosts jilted by justice. *Daily Times* [Chicago, Ill]. July 7, 1937:3; Doty M. Ottawa's doomed women. *Daily Times.* [Chicago, Ill]. March 17, 1936:3.
12. Lang D. A reporter at large: a most valuable accident. *New Yorker.* May 2, 1959:49+.

Chapter 1: Dawn of a Miracle

1. Curie M. *The Discovery of Radium.* Address by Madame M. Curie at Vassar College, May 14, 1921. Ellen S. Richards Monographs, No. 2. Poughkeepsie, NY: Vassar College;1921:5.
2. Curie M; Kellogg C, Kellogg V, trans. *Pierre Curie: With Autobiographical Notes by Marie Curie.* New York, NY: Macmillan; 1923:96.
3. Weeks ME, Leicester HM. *Discovery of the Elements.* 7th ed. Easton, Pa: *Journal of Chemical Education*; 1968:778.
4. The publication also included G. Bemont, who had collaborated with the Curies in their experiments.
5. Weeks and Leicester: 782.
6. Curie. *Pierre Curie: ...* 185.
7. Ostwald W, as quoted in Weeks and Leicester: 783.

8. Curie. *Pierre Curie:* ... 104.
9. Clarke JH, Mathur KN. *Radium as an Internal Remedy: Especially Examined in Cases of Skin Diseases and Cancer.* New Delhi, India: B. Jain; 1990:8.
10. Curie. *Pierre Curie:* ... 118.
11. Radium at $900,000 a pound. *Mineral Collector.* May 1903:44.
12. See for example DiSantis DJ, DiSantis DM. Wrong turns on radiology's road to progress. *Radiographics.* Nov. 1991:1121–1138. Also, Badash L. *Radioactivity in America: Growth and Decay of a Science.* Baltimore, Md: Johns Hopkins University Press; 1979:24–32.
13. Common sense and radium. *Med Record.* 1904; 65:780.
14. Klickstein HS. *Marie Sklodowska Curie, Recherches sur les Substances Radioactive:* A Bio-Bibliographical Study. Vol. 2. St. Louis, Mo: Mallinckrodt Chemical Works, Mallinckrodt Classics of Radiology; 1966:1.
15. Schlessinger BS, Schlessinger JH, eds. *The Who's Who of Nobel Prize Winners 1901–1990.* Phoenix, Ariz: Oryx Press; 1991:164.
16. McGrayne SB. *Nobel Prize Women in Science: Their Lives, Struggles, and Momentous Discoveries.* New York, NY: Birch Lane; 1993:28.
17. Schlessinger and Schlessinger: 6.
18. Hammer WJ. *Radium, and Other Radioactive Substances; Polonium, Actinium, and Thorium, With a Consideration of Phosphorescent and Fluorescent Substances, the Properties and Applications of Selenium and the Treatment of Diseases by the Ultra-Violet Light.* New York, NY: Van Nostrand; 1903:29.
19. Hammer: 27.
20. Curie. *The Discovery of Radium:* ... 1.
21. Reid R. *Marie Curie.* New York, NY: Saturday Review-Dutton; 1974:273.
22. Vacirca SJ. Radiation injuries before 1925. *Radiol Technol.* 1968;39:347–348.

Chapter 2: The First Nuclear Industry

1. Flannery JM, as quoted in Miller AL, Personal reminiscences of the early history of the radium extraction industry in the U.S. In: *Group Research Report: July 1965 Through June 1968.* ANL 7461. Argonne, Ill: Argonne National Laboratory; July, 1968:92.
2. Badash L. *Radioactivity in America: Growth and Decay of a Science.* Baltimore, Md: Johns Hopkins University Press; 1979:25.
3. Hammer WJ. Memorandum of William J. Hammer's connection with radium. Unpublished memorandum, n.d. Collection 69, Box 21A. Located at: William J. Hammer Collection, Archives Center, National Museum of American History, Smithsonian Institution, Washington, DC.
4. See the following Edison notebook entries N031211, N031212.1, N031212.2, N031229, N061118, N070319, N080213. Located at: Archives of the Edison National Historic Site, West Orange, NJ.
5. Radium a peril. *Penny Press* [Middletown, Conn.]. August 28, 1903;n.p.
6. Edison's friends fear radium hurt. *New York Daily News.* December 29, 1903;n.p.
7. As quoted in Radium in medicine. *New York Times.* July 17, 1904:5.
8. Jefferies TC. The story of radium in America. *Current History.* 1921;14:450; for an interesting overview of the valley's colorful past see Greger HE. *The Hell That Was Paradox.* Boulder, Co: Johnson Press; 1992.
9. Eckert JE. *An Economic Study of the Radium Industry.* [master's thesis] Cambridge, Mass.: Massachusetts Institute of Technology; 1959:49.

10. Hillebrand WF, Ransome FL. On carnotite and associated vanadiferous minerals in western Colorado. *Science.* August, 1900:120–144. Also, Mitchell RS. *Mineral Names: What Do They Mean?* New York, NY: Van Nostrand Reinhold; 1979:100.
11. Quoted from Parsons CL, Preface. In: Moore RB, Kihil KL. *A Preliminary Report on Uranium, Radium, and Vanadium.* U.S. Bureau of Mines. Bulletin 70, Mineral Technology 2. Washington, DC: U.S. Government Printing Office; 1913:8.
12. Lockwood ST, letter to Badash L. July 11, 1963.
13. Phillips AH. Radium in an American ore. *Proc Am Philos Soc.* 1904;43:160.
14. Lockwood letter.
15. First National Bank of Pittsburgh. Flannery Bolt Company. In: *The Story of Pittsburgh: Iron and Steel.* Vol. 1, No. 3. January, 1920:n. p.
16. Adventures in industry, Part 1: Vanadium. *Vancoram Review.* Spring 1956:6–9,17; see also Flannery J R. Vanadium, a romance of Catholic enterprise. In: McGuire CE, ed. *Catholic Builders of the Nation.* New York, NY: Catholic Book Company. 1935:193–200; Sketch of the life and work of Joseph M. Flannery. *Radium.* 1920;14:99–111.
17. Fleming R. Steel girders from the roof of the earth: how vanadium came to Pittsburgh. *Pittsburgh Press.* July 27, 1930:1–2.
18. Wandersee J. The reminiscences of Mr. John Wandersee. Unpublished manuscript. Located in: Ford Motor Company Archives, Oral History Section, Sept. 1952, 25. Henry Ford Museum and Greenfield Village, Dearborn, Mich.
19. Hogan WT. *Economic History of the Iron and Steel Industry in the United States.* Vol. 2, Part 3. Lexington, Mass: Lexington Books; 1971:672.
20. Miller: 91.
21. Miller: 95.
22. Kunz GF, Failla G. Radium—the supreme marvel of nature's storehouse. *Nat Hist.* 1921;21:523.
23. Ganley W. Personal communication from William Ganley, President, Radium Dial Company (New York City), former officer of Standard Chemical Company, Apr. 28, 1938. Unpublished manuscript. Located at: Pittsburgh Industries, Radium, Files, Carnegie Library, Pittsburgh, Pennsylvania: 2.
24. For a comprehensive biography of Kelly, see Davis AW. *Dr. Kelly of Hopkins: Surgeon, Scientist, Christian.* Baltimore, Md: Johns Hopkins University Press; 1959.
25. Bruyn K. *Uranium Country.* Boulder, Co: University of Colorado Press; 1955:66–74.
26. For a brief biography of Lane, see Vexler RI. *The Vice-Presidents and Cabinet Members: Biographies Arranged Chronologically by Administration.* Vol. 2. Dobbs Ferry, NY: Oceana; 1975:496–499.
27. Hearings were held in the House on bills H. J. Res. 185 and 186 from January 19–28, 1914, and in the Senate on S. 4405 from Feb. 10–24, 1914, during the 63rd Congress, Second Session; see also Hess FL. Radium, uranium, and vanadium. In: U.S. Bureau of Mines. *Mineral Resources of the United States 1914.* Washington, DC: U.S. Government Printing Office, 1916:943.
28. Lind SC. Radium production in America. *Chem Metallur Eng.* 1922:26:1012–1013; Hart SS. The Denver radium boom and the Colorado School of Mines. *Mines Mag.* February, 1986:8–9.
29. Landa ER. The first nuclear industry. *Sci Am.* November, 1982:188–189.
30. Landa ER. Buried treasure to buried waste: the rise and fall of the radium industry. *Colo School Mines Q.* 1987;82(2):25.

31. Viol CH. The story of Mme. Curie's gram of radium. *Radium.* 1921;17:37–52.
32. Ganley: Personal communication, 6.

Chapter 3: Radium Medicine

1. Proescher F. The intravenous injection of soluble salts. *Radium.* 1914;2:45–46.
2. See for example the comments of Lee CB in All radium plants in federal inquiry. *New York Times.* June 21, 1925:18.
3. Curie M. *Pierre Curie: With the Autobiographical Notes of Marie Curie.* New York, NY: Dover; 1963:56.
4. Sowers ZT. The uses of radium. *Am Med..* 1903;6:261.
5. Frame PW. Natural radioactivity in curative devices and spas. *Health Phys.* 1992;62 (suppl):S80.
6. Hammer WJ. Memorandum of William J. Hammer's connection with radium. Unpublished memorandum. Collection 69, Box 21A. Located at William J. Hammer Collection, Archives Center, National Museum of American History, Smithsonian Institution, Washington, DC.
7. Workers of the Writers' Program of the Work Projects Administration in the State of Arkansas. *Arkansas: A Guide to the State.* New York, NY: Hastings House; 1941:153159.
8. Postcard of the New Imperial Bath House, Hot Springs, Ark. Located at Curt Teich Postcard Archives, Lake County Museum, Wauconda, Ill.
9. Radium Ore Revigator Co. *The Perpetual Health Spring in Your Home.* San Francisco, Calif: Radium Ore Revigator Co.; 1925:11.
10. Frame PW. Radioactive and radium sources in medical museums. *Caduceus.* 1991: 7(2):52.
11. Frame. *Caduceus.*: 53.
12. Rowntree LG, Baetjer WA. Radium in internal medicine: Its physiologic and pharmacologic effects. *JAMA.* 1913;61:1439.
13. Gudzent F. Zur frage der vergiftung mit thorium x. *Berliner Klinishe Wochenschrift.* 1912;49:934.
14. Lohe H. Toxikologishe beobachtungen über thorium x bei mensch und tier. *Virchows Arch.* 1912;209:166.
15. Proescher F. The intravenous injection of soluble radium salts in man. *Radium.* July 1913:910.
16. Brues AM. *The radium dial.* Address before the Chicago Literary Club, January 22, 1973:19. Located at Radium Archive, Argonne National Laboratory, Argonne, Ill.
17. Field CE. Radium and researcha protest. *Med Record.* 1921;100:764.
18. Mount HA. Applying radium to cure man's ills. *Sci Am.* 1920;123:390.

Chapter 4: Mysterious Deaths

1. Blum T, as quoted in Report on health of girls employed by U.S. Radium Corporation, AC 11902, Reel No. RB-3. Located in Raymond H. Berry Papers, U.S. Library of Congress, Washington, DC.
2. Hammer WJ. Memorandum of William J. Hammer's connection with radium. Unpublished memorandum, Collection 69, Box 21A. Located at William J. Hammer Collection, Archives Center, National Museum of American History, Smithsonian Institution, Washington, DC.

3. Viol CH, Kammer GD. The application of radium in warfare. *Trans Am Electrochem Soc.* 1918;32:388–390.
4. Statement of Studer C in *Canfield v. U.S. Radium Corp.* March 6, 1930. Microfilm Drawer 4, Roll 2. Located in Radium Archive, Argonne National Laboratory, Argonne, Ill.
5. Wall FE. Sabin Arnold von Sochocky: 1882–1928. In: Miles WD, ed. *American Chemists and Chemical Engineers.* Washington, DC: American Chemical Society; 1976:488.
6. Wall: 488.
7. Enter old firm with new name: Radium Luminous Material Corporation now U.S. Radium Corporation. *Orange Advertiser* [Orange, NJ]. September 2, 1921:n.p.
8. Enter old firm....
9. Martland HS. Occupational poisoning in manufacture of luminous watch dials: general review of hazard caused by ingestion of luminous paint, with especial reference to the New Jersey cases. *JAMA.* 1929;89:466.
10. Enter old firm. ...
11. Keeney RM. Radium. In: Roush GA, Butts A, eds. *The Mineral Industry: Its Statistics, Technology and Trade During 1918.* New York, NY: McGraw-Hill; 1919:639.
12. Hess VF. Harnessing radium to motor service. Radium Archive.
13. Keeney: 639.
14. The radium watch industry. *Eng Min J.* 1920;109:882.
15. Herbert LG. Importance of radium paints. *Metal Industry.* 1921:408.
16. Herbert LG. Dickinson R. Even radium can be advertised: Undark campaign points way for basic industry. *Printers' Ink.* June 24, 1920:2528.
17. Eckert JE. *An Economic Study of the Radium Industry.* [master's thesis] Cambridge, Mass.: Massachusetts Institute of Technology; 1959:74.
18. Enter old firm. ...
19. Stewart E. Radium poisoning: industrial poisoning from radioactive substances. *Mon Labor Rev.* June 1929:51.
20. Wall: 488.
21. Von Sochocky SA. Can't you find the keyhole? *American Magazine.* January 1921:24–25.
22. Inventor poisoned by his radium paint. *New York Times.* June 8, 1928:13.
23. Stewart: 51.
24. George Stuart Willis. In: Hafner AW, Hunter, FW, Tarpey EM, eds. *Directory of Deceased American Physicians: 1804–1929.* Vol. 2. Chicago, Ill: American Medical Association; 1993:1691.
25. MacNeal WJ, Willis GS. A skin cancer following exposure to radium. *JAMA.* 1923;80:466–469.
26. MacNeal and Willis: 469.
27. Herbert: 408.
28. Craster CV, letter to Roach J. Jan. 3, 1923, AC 11902, Reel No. RB-3. Berry Papers.
29. Memorandum of defense: *La Porte v. U.S. Radium Corporation.* Point Number 6. Microfilm Drawer 4, Roll 2. Radium Archive.
30. Memorandum of defense. ... Point Number 5. Radium Archive.
31. Poisoned as they chatted merrily at their work. *New York American.* February 28, 1926: (American Weekly sec.)1.
32. Stewart: 41.
33. George AV, Gettler AO, Muller RH. Radioactive substances in a body five years after death. *Arch of Pathol.* 1929;7:397–405.

34. Walter F. Barry, Sr., D.D.S. Obituary. *J Am Dental Assoc.* 1942;29:1911–1912.
35. Ward EF. Phosphorus necrosis in the manufacture of fireworks. *J Ind Hyg.* 1928; 10:314330; for a brief history of the phosphorus poisoning, see Emsley J. The shocking history of phosphorus. *New Scientist.* 1977;74:769–772.
36. As quoted in Report on health of girls employed by U.S. Radium Corporation. Berry Papers.
37. Memorandum of Defense. ... Point Number 10. Radium Archive.
38. Craster. Berry Papers.
39. Craster. Berry Papers.
40. Erskine L, letter to Roach J. January 25, 1923, AC 11902, Reel No. RB-3. Berry Papers.
41. Stewart: 40.
42. Stewart: 41.
43. Radiation hazards of the industrial Atomic Age. National Consumers League, November 20, 1959:2, MSS. 18,509, Reel No. 84. Located at National Consumers' League Papers, U.S. Library of Congress, Washington, DC.
44. Stewart: 41.
45. Reich WT, Kahn A. *Theodor Blum, D.D.S., M.D.* New York, NY: New York Institute of Clinical Oral Pathology; 1963:48–55.
46. Stewart: 42.
47. Blum T, letter to the U.S. Radium Corporation, June 14, 1924; Roeder A, letter to Blum T, June 18, 1924, AC 11902, Reel No. RB-3. Berry Papers.
48. Death Certificate of Hazel Kuser, AC 11902, Reel No. RB-2. Berry Papers.
49. Blum T. Osteomyelitis of the mandible and maxilla. *J Am Dental Assoc.* 1924;11:805.
50. Reich and Kahn: 52.

Chapter 5: Medical Detectives and Social Activists

1. Wiley KGT, letter to Roeder A, Jan. 17, 1925, MSS 18509, Reel No. 86. Located at National Consumers' League Papers, U.S. Library of Congress, Washington, DC.
2. Arthur Roeder, industrialist, 75: ex-chairman of Colorado Fuel and Iron is dead—had headed U.S. Radium. *New York Times.* May 10, 1960:37.
3. Fisher I. *The Life Extension Institute.* New York, NY: Life Extension Institute; March 9, 1914; for a brief history of the institute, see *Executive Health Group: A Life Extension Institute Company.* New York, NY: Executive Health Group; 1996:23.
4. Singmaster JA, letter to Drinker CK, March 1, 1924. Located at Drinker Papers, Harvard University, School of Public Health, Boston, Mass.
5. Roeder A, letter to Drinker CK, Mar. 12, 1924. Drinker Papers.
6. Bowen CD. *Family Portrait.* Boston, Mass: Little, Brown, 1970.
7. Drinker CK, letter to Roeder A, March 15, 1924. Drinker Papers.
8. Drinker CK, letter to Viedt HB, April 29, 1924. Drinker Papers.
9. Castle WB, Drinker KR, Drinker CK. Necrosis of the jaw in workers employed in applying a luminous paint containing radium. *J Ind Hyg.* 1925;7:371–382.
10. Drinker CK, letter to McBride AF, June 30, 1925. Drinker Papers.
11. Roeder A, letter to Drinker CK, June 3, 1924. Drinker Papers.
12. Roeder A, letter to Drinker CK, June 13, 1924; Drinker CK, letter to Roeder A, April 15, 1925. Drinker Papers.
13. Roeder A, letter to Drinker CK, June 24, 1924. Drinker Papers.
14. Drinker CK, letter to Singmaster JA, July 14, 1924. Drinker Papers.

15. Goldmark J. *Impatient Crusader: Florence Kelley's Life Story.* Urbana, Ill: University of Illinois Press; 1953:191–192.
16. Angevine E. *History of the National Consumers' League: 1899–1979.* Washington, DC: National Consumers' League; 1979.
17. Annual report [1925] of Katherine G. T. Wiley, as quoted in MSS 18509, Reel No. 85, as National Consumers' League Papers.
18. Annual report of Wiley. National Consumers' League Papers.
19. Annual report of Wiley. National Consumers' League Papers.
20. Radium dial cases, MSS 18509, Reel No. 85. National Consumers' League Papers.
21. Radium dial cases. National Consumers' League Papers.
22. Sklar KK, ed. *The Autobiography of Florence Kelley: Notes of Sixty Years, by Florence Kelley.* Chicago, Ill: Charles H. Kerr; 1986:14.
23. Hamilton A. *Exploring the Dangerous Trades: The Autobiography of Alice Hamilton.* Boston, Mass: Little, Brown; 1943; Sicherman B. *Alice Hamilton: A Life in Letters.* Cambridge, Mass: Harvard University Press; 1984.
24. Goldmark: 197.
25. Wiley KGT notes for 19 June 1924. MSS 18509, Reel No. 86. National Consumers' League Papers.
26. Frederick Ludwig Hoffman. *National Cyclopaedia of American Biography: History of the United States.* Vol. 34. New York, NY: James T. White; 1948:66-67; Dr. Hoffman dies; actuarial expert. *New York Times.* February 25, 1946:26.
27. Statement of Hoffman FL in *Grace Fryer et al. v. U.S. Radium Corporation,* August 25, 1927. AC 11902, Reel No. RB-2. Located at Raymond H. Berry Papers, U.S. Library of Congress, Washington, DC.
28. Hamilton A, letter to Wiley KGT, February 2 and 7, 1925. MSS 18509, Reel No. 86. National Consumers' League Papers.
29. Drinker CK, letter to Roeder A, February 17, 1925. Drinker Papers.
30. Roeder A, letter to Drinker CK, April 9, 1925. Drinker Papers.
31. Hamilton A, letter to Drinker KR, April 4, 1925. Drinker Papers.
32. Drinker KR, letter to Hamilton A, April 17, 1925. Drinker Papers.
33. Drinker CK, letter to Roach J, April 22, 1925; Roach J, letter to Drinker CK, May 1, 1925; Drinker CK, letter to Roach J, May 29, 1925. Drinker Papers.
34. Hoffman FL. Radium (mesothorium) necrosis. *JAMA.* 1925;85:965.
35. Drinker CK, letter to Roeder A, June 18, 1925. Drinker Papers.
36. Stryker J, letter to Drinker CK, June 20, 1925. Drinker Papers.
37. Harrison Stanford Martland. *National Cyclopaedia of American Biography: History of the United States.* Vol. 44. New York, NY: James T. White; 1962:502–503.
38. Martland HS, letter to McBride AF, August 28, 1925. Located at Martland Papers, Special Collections, George F. Smith Library of the Health Sciences, University of Medicine and Dentistry of New Jersey, Newark, NJ.
39. Drinker CK, letter to McBride AF, June 30, 1925. Drinker Papers.
40. Reitter GS, Martland HS. Leucopenic anemia of the regenerative type due to exposure to radium and mesothorium: report of a case. *Am J Roentgenol Radium Ther.* 1926; 16:161–166.
41. Martland HS. Some unrecognized dangers in the use and handling of radioactive substances. *Proceedings of the New York Pathological Society.* 1925;25:88–92.
42. Martland HS, Conlon P, Knef JP. Some unrecognized dangers in the use and handling of radioactive substances: with especial reference to the storage of insoluble products of radium and mesothorium in the reticulo-endothelial system. *JAMA.* 1925;85:1769.

43. Fishbein M. Industrial hazards of radioactive material. *Sci Am.* 1927;136:240.
44. Frederick Flinn, physiologist, 80: former Columbia professor diestoxicologist, expert on radium poisoning. *New York Times.* April 16, 1957:33.
45. Statement of Frederick L. Hoffman in *Fryer v. U.S. Radium.*
46. Flinn FB. Radioactive material an industrial hazard? *JAMA.* 1926;87:2081.
47. Frederick B. Flinn FB, letter to Martland HS, February 11, 1929. Martland Papers.
48. Flinn FB. A case of antral sinusitis complicated by radium poisoning. *Laryngoscope.* 1927;37:347.
49. Flinn FB. Some of the newer industrial hazards. *Boston Med Surg J.* 1928;197:1312.
50. Martland HS. Occupational poisoning in manufacture of luminous watch dials. *JAMA.* 1929;89:471.
51. Kelley F, letter to Nevins A, March 25, 1929. MSS 18509, Reel No. 85. National Consumers' League Papers.
52. Martland HS. The occurrence of malignancy in radioactive persons. *Am J Cancer.* 1931;15:2507, 2513.
53. Martland. The occurrence of malignancy. ...: 2442.
54. Radium victim no. 41: lethal rays kill another of the 1920s' famous watch dial painters. *Life.* December 17, 1951:81.
55. H. S. Martland Sr., pathologist, dead: Newark ex-official, pioneer in radioactive diseases, worked at Oak Ridge. *New York Times.* May 2, 1954:88.
56. Berg S. *Harrison Stanford Martland, M.D.: The Story of a Physician, a Hospital, an Era.* New York, NY: Vantage Press; 1978:198–199.
57. Evans RD, interview with author, June 20, 1992.

Chapter 6: In Search of Justice

1. Lippmann W. Five women doomed to die. *World* [New York, NY]. May 10, 1928:14.
2. Dial painter sues concern claiming injury to health. *Newark Evening News.* March 9, 1925:6.
3. Wiley KGT, letter to Hamilton A, March 10, 1925. Located at Drinker Papers, Harvard University School of Public Health, Boston, Mass.
4. Wiley KGT, letter to Hamilton A, May 7, 1926. MSS 18508, Reel No. 86. Located at National Consumers' League Papers, U.S. Library of Congress, Washington, DC.
5. Edmonds DS, Affidavit of conference with Knef JP of Newark, NJ, at office of the U.S. Radium Corporation on May 19, 1926. AC 11902, Reel No. RB-3. Located at Raymond H. Berry Papers, U.S. Library of Congress, Washington, DC.
6. Cross WR, Affidavit of conference with Knef JP. Berry Papers.
7. Cross: 3. Berry Papers.
8. Lee CB, Affidavit of conference with Knef JP. Berry Papers.
9. Roeder A, Affidavit of conference with Knef JP. Berry Papers.
10. Heazelton FK, letter to Berry RH, June 16, 1927. AC 11902, Reel No. RB-3. Berry Papers.
11. Dr. Joseph P. Knef, dentist, dies at 76. *Newark Evening News.* October 9, 1946:30.
12. Hoffman FL, letter to Kelley F, 10 June 1926. MSS 18509, Reel No. 86. National Consumers' League Papers.
13. Arthur Roeder, industrialist, 75: ex-chairman of Colorado Fuel and Iron is dead—had headed U.S. Radium. *New York Times.* May 10, 1960:37.
14. Grace Fryer Affidavit. *Grace Fryer et al. v. U.S. Radium Corporation.* AC 11902, Reel No. RB-1. Berry Papers.

15. Quinta McDonald Affidavit. *Fryer et al. v. U.S. Radium.* Berry Papers.
16. Albina Larice Affidavit. *Fryer et al. v. U.S. Radium.* Berry Papers.
17. Edna Hussman Affidavit. *Fryer et al. v. U.S. Radium.* Berry Papers.
18. Katherine Schaub Affidavit. *Fryer et al. v. U.S. Radium.* Berry Papers.
19. Raymond H. Berry File. Located at: Alumni Records Office, Yale University, New Haven, Conn.
20. Katherine Schaub. *Radium. Survey Graphic.* 1932;21:139.
21. *Fryer et al. v. U.S. Radium.* Berry Papers.
22. New issue raised in radium poisoning. *New York Times.* July 19, 1927:25.
23. Body exhumation may decide suit: woman's corpse will be examined for evidence in $1,250,000 action against radium company. *Newark Star-Eagle.* October 10, 1927: n.p.; Exhume girl's body to find death cause: plaintiffs in suit have remains dug up in evidence hunt. *Newark Sunday Call.* October 16, 1927:n.p.
24. George AV, Gettler AO, Muller RH. Radioactive substances in a body five years after death. *Arch of Pathol.* 1929;7:397–405.
25. Hearing transcript, *Grace Fryer et al. v. U.S. Radium.* Reel RB-1. Berry Papers.
26. Hearing transcript, April 25, 1928, RB-2. Berry Papers.
27. Hearing transcript, April 26, 1928, RB-2. Berry Papers.
28. Hearing transcript, April 27, 1928, RB-2. Berry Papers.
29. Doomed to die, tell how they'd spend fortune. *Newark Sunday Call.* May 13, 1928:2.
30. Radium case off till fall. *Newark Evening News.* April 27, 1928:n.p.; Five poisoned women face court delay. *New York Times.* April 28, 1928:34; Suit of five women facing death adjourned to seek evidence. *World* [New York, NY]. April 28, 1928:n.p.
31. Steel R. *Walter Lippmann and the American Century.* Boston, Mass: Atlantic-Little, Brown, 1980; see also Mainstream media: the radium girls. In: Neuzil M, Kovarik W. *Mass Media and Environmental Conflict: America's Green Crusades.* Thousand Oaks, Calif: Sage, 1996:33-52.
32. Lippmann: 14.
33. Dayton D. Girls poisoned with radium not necessarily doomed to die: Dr. Flinn of College of Physicians and Surgeons of Columbia takes hopeful view of five New Jersey victims. *Newark Evening News.* May 11, 1928:n.p.; Five radium victim may live, he finds. *World* [New York, NY]. May 18, 1928:n.p.; Radium victim's doom denied by doctor. *Newark Star-Eagle.* May 18, 1928:n.p.
34. Lippmann W. The case of the five women. *World* [New York, NY]. May 19, 1928:n.p.
35. Suing women may survive trial in fall, but their days are few he claims: Dr. Frederick L. Hoffman of Wellesley Hills, Mass., first to discover effects of radium on workers, disputes co. doctor. *Newark Ledger.* May 24, 1928:n.p.
36. Radium poison hopeless: Mme. Curie holds out scant hope for girls facing radium death. *New York Evening Journal.* May 26, 1928:1.
37. Radium poison hopeless....: 2.
38. Radium cases socialist topic: Norman Thomas sees example of capitalist system in women victims. *Newark Evening News.* May 21, 1928:n.p.
39. Endeavor to speed radium suits fails: Backes holds chancery powerless to advance case, but suggests defense submit it for ruling. *Newark Star-Eagle.* May 22, 1928:n.p.; Seeks trial Monday of five radium suits: counsel of women who ask $1,250,000 as poisoning victims to have circuit court hearing. *New York Times.* May 25, 1928:n.p.
40. Judge William Clark. In: Winfield Scott Downs, ed. *Encyclopedia of American Biography.* New Series. Vol. 29. New York, NY: American Historical; 1959:241–243.

41. U.S. judge offers to be mediator in 'ray paint' suits. *World* [New York, NY]. May 30, 1928:n.p.
42. Radium victims win $50,000 and pensions in suit settlement. *New York Times.* June 5, 1928:1.
43. Terms displease victim of radium: Mrs. McDonald disappointed over 'small sum' promised by the Corporation. *Daily Courier* [Orange, NJ]. June 5, 1928:n.p.
44. One defendant fails to agree to radium pact. Unknown newspaper, June 5, 1928:4. Located at Radium Archive, Argonne National Laboratory, Argonne, Ill.
45. Lee CB, letter to Harris LI, June 18, 1928. Radium Archive.
46. Poison paintbrush. *Time.* June 4, 1928:41.
47. Whose responsibility? *Sci Am.* 1928;139:108.
48. The new death. *Daily Courier* [Orange, NJ]. May 3, 1928:n.p.
49. All radium plants in federal inquiry: labor department starts search to discover the cause of 'radium necrosis.' *New York Times.* June 6, 1925:18; Effects of use of radioactive substances on the health of workers. *Mon Labor Rev.* May 1926:18; Stewart E. Transcript of U.S. Public Health Conference on Radium in Industry, December 20, 1928, Washington, DC. MSS 18509, Reel No. 85. National Consumers' League Papers.
50. Federal investigation of radium poisoning asked by civic groups: appeal to surgeon general to protect workers from slow death. *World* [New York, NY]. July 15, 1928:n.p.; Cumming HS, letter to Hamilton A, July 18, 1928. MSS 18509, Reel No. 85. National Consumers' League.
51. Stewart E. Transcript of U.S. Public Health Conference: 24, 26. National Consumers' League Papers.
52. Moore RB. Transcript of U.S. Public Health Conference: 27. National Consumers' League Papers.
53. Martland HS. Occupational poisoning in luminous watch dials: general review of hazard caused by ingestion of luminous paint, with especial reference to the New Jersey cases. *JAMA.* 1929;89:558.
54. Martin RE. Doomed to die—and they live! Medical science brings a ray of hope into the tragic lives of five girls poisoned by radium. *Popular Science.* July 1929: 17-19, 136.
55. Schaub. Radium: 140–141.
56. Autopsy performed on radium victim: tumors found in body of Mrs. McDonald similar to those in four other cases. *New York Times.* December 9, 1929:22.
57. Radium worker dies: second in five women who sued in New Jersey succumbs. *New York Times.* February 19, 1933:31; Twenty-second worker dies of radium poison: woman became ill 14 years ago painting watch dials in New Jersey plant. *New York Times.* October 28, 1933:17.
58. Radium claims its 28th victim: Mrs. Edna Hussman was one of five given year to live 11 years ago. Unknown newspaper, March 31, 1939:n.p. Radium Archive.
59. Last radium victim dies: 28th to succumb from malady contracted almost 25 years ago. *New York Times.* November 19, 1946:7.
60. *La Porte v. U.S. Radium Corporation*, 13 F. Supp. (1935) 263–277.
61. *La Porte v. U.S. Radium*, 264–265.
62. *La Porte v. U.S. Radium*, 267–268, 270–271, 277.
63. Berg S. *Harrison Stanford Martland, M.D.: The Story of a Physician, a Hospital, an Era.* New York, NY: Vantage Press; 1978:187.
64. A list of all known deaths can be found in the Appendix Table A1.

Chapter 7: The Ottawa Society of the Living Dead

1. Doty M. Kin reveal agony of radium victims. *Daily Times* [Chicago]. March 18, 1936:3–4.
2. Swen Kjaer File, Bureau of Labor Statistics, U.S. Department of Labor. Unpublished report on the bureau's investigation of the nation's radium industry 1925–1929. Located in Special Collections, George F. Smith Library of the Health Sciences, University of Medicine and Dentistry of New Jersey, Newark, NJ.
3. Mullner, R. *The Ottawa dial workers*. Unpublished manuscript. 1995:5.
4. Miller CE. Radium in humans. *Radiological Physics Division Semiannual Report:* January-June 1957. Proceeding Report: ANL-5679. Argonne, Ill: Argonne National Laboratory, 1958;58.
5. Transcript of testimony of Catherine W. Donohue in *Catherine W. Donohue v. Radium Dial Co.*, Illinois Industrial Commission, February 10–11, 1938, Case No. 212150. Located at Radium Archive, Argonne National Laboratory, Argonne, Ill.
6. Brues AM. *The radium dial.* Paper presented at the Chicago Literary Club, January 22, 1973:22. Radium Archive.
7. Mullner: 10.
8. Miller: 58.
9. Kjaer File: n.p.
10. Gosling FG. Dial painters project: Argonne National Laboratory's documentation of radium hazards to workers. *Labor's Heritage*. 1992:4(2):67.
11. Kjaer File.
12. Kjaer File.
13. Mullner: 21.
14. Stewart E. Radium poisoning: industrial poisoning from radioactive substances. *Mon Labor Rev.* June 1929:46.
15. *Donohue v. Radium Dial Co.*: 33.
16. Mullner: 23.
17. *Donohue v. Radium Dial Co.*: 33.
18. Radium victims win $50,000 and pension in suit settlement. *New York Times*. June 5, 1928: 1.
19. Statement by the Radium Dial Company. *Daily Republican-Times* [Ottawa, Ill]. June 7, 1928: 3.
20. Keane AT, Lucas HF, Markun, F, Essling, MA., Holtzman, RB. The estimation and potential radiobiological significance of the intake of 228-ra by early ra dial workers in Illinois. *Health Phys.* 1986;51:313–327.
21. Hoffman FL. Radium (mesothorium) necrosis. *JAMA*. 1925;85:961–965.
22. Martland HS, Conlon P, Knef JP. Some unrecognized dangers in the use and handling of radioactive substances: with especial reference to the storage of insoluble products of radium and mesothorium in the reticulo-endothelial system. *JAMA*. 1925;85:1769–1776.
23. Stewart: 1233.
24. Doty: 4.
25. Unidentified newspaper article. Radium Archive.
26. Stewart: 1239.
27. Edith Schomas, interview with the author. April 1991.
28. Mullner: 33.
29. Ridings J. Check radium victims: body exhumed here. *Daily Times* [Ottawa, IL]. May 19, 1978:5.

30. Mullner: 16
31. Doty: 3-4.
32. *Memorandum: general considerations. U.S. Federal Bureau of Investigation report of activities of the Union Miniere du Haut Katanga in the United States*, 9 Dec. 1941. Manhattan Engineer District Investigation Files, RG77, Entry 8, Box 101. Located at National Archives, Washington, DC.
33. Main J. Fifteen Walking Ghosts Jilted by Justice. *Daily Times* [Chicago, Ill]. July 7, 1937: 3.
34. *Donohue v. Radium Dial Co.*: 54.
35. *Inez Corcoran Vallat v. Radium Dial Co.*, 360 Ill. 407, 196 N.E. 485.
36. Griffin F. Society of the living dead! *Star Weekly* [Toronto, Can.]. April 23, 1938:3.
37. *Vallat v. Radium Dial Co.*: 407.
38. Mullner: 23.
39. Main: 3.
40. Snider AJ. Ranks of 'living dead' dwindle in 25 years: victims of radium poisoning. *Chicago Daily News*. June 13, 1953:6.
41. Sholis V. Ottawa radium firm now in N.Y. *Daily Times* [Chicago, Ill]. July 8, 1937:3.
42. *Donohue v. Radium Dial Co.*: 119.
43. *Donohue v. Radium Dial Co.*: 24.
44. *Donohue v. Radium Dial Co.*: 34–35.
45. Gardner V. Brief upholds woman worker's radium claims: negligent poisoning is charged by lawyer. *Chicago Tribune*. February 27, 1938:10.
46. Mullner: 40.
47. *Donohue v. Radium Dial Co.*: 50.
48. Wins radium poison suit: woman gets cash, compensation and pension award. *New York Times*. April 6, 1938:29.
49. Walsh K. 'Pray for me,' radium victim's plea. *Daily Times* [Chicago, Ill]. June 24, 1938:3.
50. Walsh K. Pray for 'living death' victim. *Daily Times* [Chicago, Ill]. June 26, 1938:5.
51. Inquest jury votes radium poison verdict. *Daily Republican-Times* [Ottawa, Ill]. July 29, 1938:1, 16.
52. Eleventh victim taken by radium. *Chicago American*. July 27, 1938:3.
53. *Cases Argued and Decided in the Supreme Court of the United States: October Term, 1939*, in 308, 309, 310 U.S. Book 84 Lawyers' Edition. Rochester, NY: Lawyers Cooperative; 1940.
54. Court upholds award in girl's radium death. *Chicago Tribune*. October 10, 1939:14.
55. A list of all known deaths can be found in the Appendix Table A1.

Chapter 8: The National Radium Scandal

1. Lain ES. The present status of radium therapy. *South Med J*. 1922;15:495.
2. Martland HS, Conlon P, Knef JP. Some unrecognized dangers in the use and handling of radioactive substances: with especial reference to the storage of insoluble products of radium and mesothorium in the reticulo-endothelial system. *JAMA*. 1925;85:1773.
3. Martland HS, letter to Carter HA. July 13, 1932. Located at Martland Papers, George F. Smith Library, University of Medicine and Dentistry of New Jersey, Newark, NJ.
4. Allen EV, Bowing HH, Rowntree LG. The use of radium in internal medicine: further experience. *JAMA*. 1927;88:164–168.
5. Schlundt H, Nerancy JT, Morris JP. The detection and estimation of radium in living persons: IV. The retention of soluble radium salts administered intravenously. *Am J*

Roetgenol Radium Ther. 1933;30:515–522; Finkel AJ, Miller CE, Hasterlik RJ. Radium-induced malignant tumors in man. In: Mays CW, Jee, WSS, Lloyd, RD, Stover, BJ, Dougherty, JH, Taylor, ed. *Delayed Effects of Bone-Seeking Radionuclides.* Salt Lake City, Utah: University of Utah Press; 1969:210-211.

6. Hold auto sellers for $750,000 fraud. *New York Times.* May 8, 1915:24; Put auto scheme under fraud order. *New York Times.* September 12, 1915(sec. 2):8.
7. Cramp AJ. *Nostrums and Quackery.* Vol. 2. Chicago, Ill: American Medical Association; 1921:603.
8. Bureau of Investigation, American Medical Association. Radium as a 'patent medicine:' the methods and activities of William J. A. Bailey in the field of radioactivity. *JAMA.* 1932;98:1397–1399.
9. Bailey Radium Laboratories. *Modern Treatment of Arthritis and Kindred Conditions with* Radium Water. East Orange, NJ: Bailey Radium Laboratories; 1927:30. Radium Cures Folder No. 0721-01, Located at American Medical Association Library, Chicago, Ill.
10. Federal Trade Commission Complaint Against Bailey Radium Laboratories, Inc., et al. Docket 1756. *Federal Trade Commission Decisions; Findings, Orders, and Stipulations.* Vol. 15. Washington, DC: U.S. Government Printing Office; 1933:419–429.
11. Doubts that radium caused seven deaths: director of Bailey Laboratories points to Mme. Curie's health as example. *New York Times.* June 22, 1925: 17.
12. Bailey WJA, letter to the editor entitled Radium deaths, June 20, 1925. Located at Radium Archive, Argonne National Laboratory, Argonne, Ill.
13. Rowland RE. *Radium in Humans: A Review of U.S. Studies.* ANL/ER-3 UC-408. Argonne, Ill: Argonne National Laboratory; 1994:8.
14. Evans RD. The effects of skeletally deposited alpha-ray emitters in man. *Br J Radiol.* 1966;39:882.
15. Eben M. Byers funeral here: local manufacturer, sportsman dies of radium poisoning. *Pittsburgh Post-Gazette.* April 1, 1932:1,4; Eben M. Byers, radium water drinker, dies of *Poisoning. New York Herald Tribune.* April 2, 1932:n.p.
16. Testimony of Eben Byers on 10 Sept. 1931. Federal Trade Commission Docketed Case Files, Docket 1756, *FTC v. Bailey Radium Laboratories*, 491. RG122. Located at National Archives, Suitland, Md.
17. Macklis RM. The great radium scandal. *Sci Am.* August 1993:95.
18. Doolittle W. AEC to probe 1931 death linked to Jersey 'water.' *Daily Journal* [Elizabeth, NJ]. March 18, 1964:n.p.
19. Federal Trade Commission Decisions: 419.
20. Robert W. Winn as quoted in Radium drinks. *Time.* April 11, 1932:48.
21. Eben M. Byers dies of radium poisoning. *New York Times.* April 1, 1932:1,11.
22. Death of Byers stirs inquiry on radium 'cures.' *New York Herald Tribune.* April 2, 1932: n.p.
23. Bureau of Investigation: 1397–1399.
24. The great radium mystery. *Fortune.* February 1934:105.

Chapter 9: Safety Standards and the Atomic Bomb

1. Seaborg GT as quoted in Kathren RL, Gough JB, Benefiel GT, eds. *The Plutonium Story: The Journals of Professor Glenn T. Seaborg, 1939-1946.* Columbus, Ohio: Battelle Press; 1994:378.
2. Radium and aviation. *New York Times.* November 26, 1939; B7.

3. Morse, KM, Kronenberg MH. Radium painting: hazards and precautions. *Ind Med.* 1943;12:810.
4. Interview of Robley D. Evans by Lauriston S. Taylor on April 20, 1978. In: Taylor LS, Sauer KG. *Vignettes of Early Radiation Workers (Transcripts of the Videotape Series).* Rockville, Md: U.S. Department of Health and Human Services; 1984:67–81.
5. Evans RD. Radium poisoning: a review of present knowledge. *Am J Public Health.* 1933;23:1017-1023.
6. For a history of whole-body counting see Medical and Health Sciences Division, Oak Ridge Associated Universities. *Current Status of Whole-Body Counting to Detect and Quantify Previous Exposure to Radioactive Materials.* Report to the National Cancer Institute. Oak Ridge, Tenn: Oak Ridge Associated Universities; 1987.
7. Evans RD. Inception of standards for internal emitters, radon and radium. *Health Phys.* 1981;41:437–448.
8. Evans. Inception of standards ... .: 443.
9. U.S. Department of Commerce, National Bureau of Standards. *Safe Handling of Radioactive Luminous Compound.* National Bureau of Standards Handbook H27. Washington, DC: U.S. Government Printing Office; 1941.
10. Pratt RM. Review of radium hazards and regulation of radium in industry. *Environment International.* 1993;19(special issue):481.
11. Text of statement by Truman, Stimson on development of atomic bomb. *New York Times.* August 7, 1945:4.
12. Gosling FG. *The Manhattan Project: Science in the Second World War.* Energy History Series. Washington, DC: U.S. Department of Energy, 1990.
13. Hacker BC. *The Dragon's Tail: Radiation Safety in the Manhattan Project, 1942-1946.* Berkeley, Calif: University of California Press; 1987:53.
14. Seaborg GT, letter to the author, November 3, 1997.
15. Seaborg, November 3, 1997.
16. A description of the radium plant can be found in Evans RD. Protection of radium dial workers and radiologists from injury by radium. *J Ind Hyg Toxicol.* 1943;25:253–274.
17. The description of the explosion and its effects is from Farrell TF as quoted in General Groves's Report on 'Trinity,' July 18, 1945. In: Cantelon PL, Hewlett RG, Williams RC, eds. *The American Atom: A Documentary History of Nuclear Policies From the Discovery of Fission to the Present.* 2nd ed. Philadelphia, Pa: University of Pennsylvania Press; 1991:51–59.
18. General Groves's report ... : 53.
19. Gosling: 48.
20. The blasting power of the new bomb. *New York Times.* August 7, 1945:3.
21. Gosling: 51.
22. Text of statement by Truman ... : 4.
23. Hiroshima a 'city of dead.' *New York Times.* August 9, 1945:6.
24. Gosling: 54.
25. Cantril ST, Parker HM. Status of health and protection at the Hanford Engineer Works. In: Stone RS, ed. *Industrial Medicine on the Plutonium Project: Survey and Collected Papers.* National Nuclear Energy Series: Manhattan Project Technical Section. Division IV—Plutonium Project Record. Vol. 20. New York, NY: McGraw-Hill; 1951:478.
26. Lang D. A reporter at large: a most valuable accident. *New Yorker.* May 2, 1959: 49+.

27. Merril Eisenbud as quoted in Lang: 52.
28. Eisenbud in Land: 52.

Chapter 10: Under Radioactive Clouds

1. Plumb RK. Radium victims sought in study. *New York Times*. October 9, 1958:39.
2. See for example Dietz D. *Atomic Energy in the Coming Era*. New York, NY: Dodd, Mead, 1945; Wendt G, Geddes DP, eds. *The Atomic Age Opens*. New York, NY: World, 1945; Del Sesto SL. Wasn't the future of nuclear energy wonderful? In: Corn JJ, ed. *Imagining Tomorrow: History, Technology, and the American Future*. Cambridge, Mass: MIT Press; 1986:58–76.
3. Martland HS. *Collection of Reprints on Radium Poisoning: 1925-1939*. Oak Ridge, Tenn: Technical Information Service, U.S. Atomic Energy Commission; 1951.
4. For a comprehensive history of the ABCC (now called the Radiation Effects Research Foundation) see Schull WJ. *Effects of Atomic Radiation: A Half-Century of Studies From Hiroshima and Nagasaki*. New York, NY: Wiley; 1995.
5. Leviero A. Truman orders hydrogen bomb built for security pending an atomic pact; Congress hails step; board begins job. *New York Times*. February 1, 1950:1, 3.
6. Trussell CP. Atom bomb testing ground will be created in Nevada. *New York Times*. January 12, 1951:1, 12.
7. Atomic test blast shakes Las Vegas, fifty miles away. *New York Times*. January 28, 1951: 1, 38.
8. Radioactivity up in Chicago. *New York Times*. November 3, 1951:6.
9. Plumb RK. Increased radiation found in East; laid to atom tests, held harmless. *New York Times*. February 3, 1951:1, 5.
10. Spoiled paper. *New York Times*. October 26, 1952:D10.
11. Seaborg GT. *Kennedy, Khrushchev, and the Test Ban*. Berkeley, Calif: University of California Press; 1981.
12. See for example Loutit JF. *Irradiation of Mice and Men*. Chicago, Ill: University of Chicago Press, 1962.
13. Leary WE. In 1950's, U.S. collected human tissue to monitor atomic tests. *New York Times*. June 21, 1995; B8.
14. Lapp RE. *The Voyage of the Lucky Dragon*. New York, NY: Harper; 1958.
15. As quoted in Wasserman H. *Killing Our Own: The Disaster of America's Experience with Atomic Radiation*. New York, NY: Delacorte; 1982:88.
16. Pauling L. *No More War!* New York, NY: Dodd, Mead; 1958.
17. Libby WF. How dangerous is radioactive fallout? *For Pol Bull*. 1957;36:151.
18. Text of address by Stevenson calling for world pact to end hydrogen bomb tests. *New York Times*. October 16, 1956:18.
19. Clark SD. Where are the cases of radium poisoning? A plea for assistance. *JAMA*. 1958;168:761.
20. Eisenbud as quoted in Lang: 52.
21. Eisenbud in Lang: 50.
22. Plumb: 39.
23. Sharpe WD. The New Jersey radium dial workers: seventy-five years later. In: Sheehan HE, Wedeen RP, eds. *Toxic Circles: Environmental Hazards From the Workplace into the Community*. New Brunswick, NJ: Rutgers University Press; 1993:138–167.
24. Rowland RE. *Radium in Humans: A Review of U.S. Studies*. ANL/ER-3, UC408. Argonne, Ill: Argonne National Laboratory; 1994:35–51.

25. Argonne National Laboratory. *Frontiers: Research Highlights, 1946-1996.* Argonne, Ill: Argonne National Laboratory; 1997:42.
26. Snider AJ. One-hundred-thousand used as radiation guinea pigs. *Chicago Daily News.* June 9, 1958:n.p.
27. Rowland: 31–35.
28. Evans RD. *Comments on a National Center of Human Radiobiology.* December 1967. Submitted to the Division of Biology and Medicine of the U.S. Atomic Energy Commission. Located at Radium Archive, Argonne National Laboratory, Argonne, Ill.
29. Rowland: 67.
30. Valeo T. The plight of the radium people. *Herald* [Arlington Heights, Ill]. December 7, 1980(Panorama sec.):5.
31. Rowland: 72.
32. See for example Adams EE, Brues AM. Breast cancer in female radium dial workers first employed before 1930. *J Occup Med.* 1990;22:583–587; Polednak AP, Stehney AF, Rowland RE. Mortality among women first employed before 1930 in the U.S. radium dial-painting industry. *Am J Epidemiol.* 1978;107:179–196; Spiers FW. Leukemia incidence in the U.S. dial workers. *Health Phys.* 1983;44(suppl):65–72; Stebbings JH, Lucas HG, Stehney AF. Mortality from cancers of major sites in female radium dial workers. *Am J Ind Med.* 1984;5:145–148; and Thomas RG. Tumorgenesis in the U.S. radium luminizers: how unsafe was this occupation? In: van Kaick G., Karaoglou A, Kellerer AM, eds. *Health Effects of Internally Deposited Radionuclides: Emphasis on Radium and Thorium.* River Edge, NJ: World Scientific; 1995:145–148.
33. See for example Schlenker RA. Radioactivity in persons exposed to fallout from the Chernobyl reactor accident. In: D. Hemphill, ed. *Trace Substances in Environmental Health—XXI.* Columbia, Mo: University Missouri Press; 1987:213–218; Rundo J. Some aspects of radon and its daughter-products in man and his environment. In: Vohra, KG, ed. *Natural Radiation.* New Delhi, India: Wiley; 1982:155–162; Rundo J, Toohey RE. Radon in homes and other technologically enhanced radioactivity. *Environmental Radioactivity.* Bethesda, Md: National Council on Radiation Protection and Measurement; 1983:17–26; Rundo J, Markun F, Plondke NJ. On the exhalation rate of radon by man. In: *Indoor Radon and Lung Cancer: Reality or Myth?* Columbus, OH: Battelle Press; 1992:253–262.
34. Rowland: 114.
35. *Internal Emitter Close-Out Plan, April 21, 1994.* Unpublished report. Radium Archive.

Chapter 11: Conclusion

1. Roush W. Learning from technological disasters. *Technol Rev.* 1993;96(6):56.
2. Moeller DW. *Environmental Health.* Cambridge, Mass: Harvard University Press; 1992: 212.
3. Joel LG. *Every Employee's Guide to the Law.* New York, NY: Pantheon; 1993:182–204.
4. U.S. Dept of Health and Human Services, National Institute for Occupational Safety and Health. *National Occupational Research Agenda.* Washington, DC: U.S. Government Printing Office; 1996:iii.
5. *National Occupational Research Agenda*: vii.
6. *National Occupational Research Agenda*: 38.
7. See for example White L. *Human Debris: The Injured Worker in America.* New York, NY: Seaview/Putnam, 1983:79–87 and Nelkin D, Brown MS. *Workers at Risk: Voices from the Workplace.* Chicago, Ill: University of Chicago Press; 1984:136–138, 143.

8. The estimates for the servicemen were provided by Farber S, letter to the author, November 11, 1997; estimates for the civilian population are from Mellinger-Birdsong AK. *Otolaryngology—Head and Neck Surgery.* 1996;115:429–432.
9. Finkel M. A radioactive lourdes. *New York Times Magazine.* December 24, 1995:34–35.
10. For a history of the contamination caused by the U.S. Radium Corporation of Orange, New Jersey, see Cole LA. *Element of Risk: The Politics of Radon.* Washington, DC: AAAS Press; 1993.
11. For the full details of the proposed cleanup, see the U.S. Environmental Protection Agency's Record of Decision ID No. NJD980785653, Montclair/West Orange, NJ, and ID No. NJD980785646, Glen Ridge, NJ.
12. Galant D. Living with a radium nightmare. *New York Times.* September 29, 1996;13:10.
13. At the end of 1997, I called the Waterbury public health department to see if any radium contamination had been found at the former Waterbury Clock Company. Although the department was aware of the radium dial tragedy, they were unaware that it had taken place in their city. Later, I contacted the Connecticut Department of Environmental Protection and asked them to conduct an investigation. In early 1998, they found one of the buildings to be contaminated.
14. See for example Glendinning C. *When Technology Wounds: The Human Consequences of Progress.* New York, NY: Morrow; 1990.

Index

Academy of Sciences of Vienna, 10
Addams, Jane, 61
Aircraft instrument dials, 1, 119
Alamogordo bomb test, 126
Allen, George Herbert, 50
Alpha radiation, 125, 130
American Cancer Society, 63
American Journal of Cancer, 72
American Journal of Public Health, 121
American Linseed Oil Company, 77
American Medical Association, 38, 65, 69, 118
   New and Nonofficial Medical Remedies, 38, 118
American Medicine, 33
American Statistical Association, 63
Andrews, John B., 62
Anemia, 1, 12, 38, 98
   Aplastic, 5, 13, 46
Antibiotics, 157
Argonne National Laboratory, vii, 134-138
   Morgue, 137
   Whole-body counter or iron room, 137
Arium, 112
Arms race, 131-133
Atomic bomb, 123-125
Atomic Bomb Casualty Commission (ABCC), 130
Atomic bomb survivors, 6, 130
Atomic energy, 129
Atomic-powered airplanes, 129
Atomic rockets, 129
Austrian government, 9, 18

Bad Gastein, Austria, 34
Baetjer, W. A., 36
Bailey, Frederick, 110
Bailey, William J. A., 110-114
Bale, Tony, 1
Barker, George F., 15, 31
Barry, Walter F., 49–50, 58, 61

Bath, England, 34
Becquerel, Henri, 7, 10, 11, 17
Belgian Congo, 28
Belgium, 28
Bell, Alexander Graham, 31
Bellevue Hospital, 67
Berry, Raymond H., 80–85
Blum, Theodor, 41, 51-53, 61
Bohemia, 9, 18
Bridgeville, Pennsylvania, 22
Buffalo, New York, 20-21
Byers, Eben MacBurney, 114-118, 120, 140

California Institute of Technology (Caltech), 120
Cambridge, England, 33
Cancer, 5, 24, 31, 38, 46, 140
Canonsburg, Pennsylvania, 25, 26
Carlough, Margaret, 68, 75
Carnegie Engineering Corporation, 110
Carnot, Marie Adolphe, 19
Carnotite ore, 18-21, 24, 27, 34
Castle, William B., 58
Center for Human Radiobiology, vii-viii, 137-138
Chernobyl reactor accident, 138
Chicago, 34, 101-102, 105, 107, 132, 137
Chicago *Daily Times*, 102
Chicago Hull House, 4, 61
Children, 60
China, 130
Clark, William, 85
Cold War, 6, 127, 130
Colorado, 18-19, 24-25, 27
Colorado Fuel and Iron Company, 77
Columbia University, 67
   Institute of Public Health, 69, 116
Cook, J. S., 101
Cornell University, 55
Craster, Charles V., 50, 68
Crookes' tube, 32

Cruse, Mary Ellen, 96-97
Cumming, Hugh S., 87
Curie, Marie, 4, 7-13, 21, 28, 83
Curie, Pierre, 7-13, 21, 31
Czech Republic, 9
Czechoslovakia, 123, 130

Dalitsch, Walter W., 105
Darrow, Clarence, 102
Davidson, James B., 49-50, 58, 61
Dentists, 2, 48-49, 64
Denver, 27, 35
Dolores River, 18
Donohue, Catherine Wolfe, 103-108
Doty, Mary, 91
Douglas, James, 27
Drinker, Cecil K., 56-60, 61, 65-66
Drinker, Katherine R., 58, 65
Drinker, Philip, 57
Drinker Respirator or iron lung, 57

East Germany, 130
East Orange, New Jersey, 112
Eastman Kodak plant, 132
Edison, Thomas A., 15-17
Eisenhower, Dwight D., 134
Elgin State Hospital, Illinois, 110, 135
England, 41
Essex County, New Jersey, 67
Evans, Robley D., 119-122, 125, 134, 136-137

FBI (Federal Bureau of Investigation), 61
Federal Trade Commission (FTC), 117
Fernandini, Eulogio, 23
Field, C. Everett, 38
Fishbein, Morris, 69
Flannery Bolt Company, 22
Flannery, James J., 21-23
Flannery, Joseph M., 15, 21-27, 28
Flinn, Frederick B., 69-72, 81, 83, 87, 116, 122
Flower Hospital, New York City, 18
Ford, Henry, 23
Fordyce, Rufus, 105
France, 31, 33
French Academy of Sciences, 19
Fryer, Grace, 77-79, 80, 88

Gamma radiation, 69, 121, 130, 141
Ganley, William, 103
General Memorial Hospital, New York City, 27

Germany, vii, 27, 38, 41
Glen Ridge, New Jersey, 141
Great Depression, 114, 119
Grossman, Leonard J., 102-108

Hahn, Otto, 123
Hamilton, Alice, 61-62, 65-66, 83, 87
Hammer, William J., 15-16, 33, 41
Hanford Engineer Works, 126-127
Hanford, Washington, nuclear facility, 142
Harding, Warren G., 28, 29
Harvard University, 56, 61, 110
    Medical School, 66
    School of Public Health, 57, 58, 61
Health physics (radiation safety), ii, 6
Hill, Mary Jennings, 116
Hiroshima, 126, 130, 131
Hoffman, Frederick L., 63-66, 70, 77, 80, 83, 96
Hot Springs, Arkansas, 34
Hussman, Edna, 77, 79, 88
Hydrogen (thermonuclear) bomb, 131

Illinois Industrial Commission, 97, 99, 102, 108
Illinois Occupational Diseases Act, 101, 102
Illinois Supreme Court, 102, 105, 108
Ingersoll, Charles, 55
Ingersoll, Robert, and Brother Company, 55
Ingersoll Watch Company, 42

Jachymov, see St. Joachimsthal
Jefferson Medical College, Philadelphia, 114
Johns Hopkins University, 27, 36
Journal of Industrial Hygiene, 66
Journal of the American Dental Association, 53
Journal of the American Medical Association, 36, 46, 69-70, 96, 109

Katanga, Africa, 28
Keane, James, 107
Kelley, Florence, 4, 61-62, 72, 87
Kelly, Howard A., 27
Kelly, Joseph A. Sr., 101, 105
Kenton, Pauline, 100
Knef, Joseph P., 48-49, 61, 75, 76-77
Kunz, George F., 41
Kuser, Hazel Vincent, 51-53
Kuser, Theodore, 52

Lackawanna, New York, 21
La Guardia, Fiorello, 72

Index

Lain, Everett S., 109
Lane, Franklin K., 27
La Porte, Irene Corby, 88-89
La Porte, Vincent, 88-89
La Porte v. U.S. Radium Corporation, 88-90
Larice, Albina Maggia, 47, 77, 79-81, 88
La Salle County Circuit Court, Illinois, 108
Las Vegas, 131
Lee, Clarence B., 85
Legal Aid Society of Newark, 61
Leman, Edwin D., 68
Leukemia, 138
Libby, Willard, 133
Life Extension Institute, 55-56, 58
Life magazine, 72
Lifetime Radium Vitalizer, 35
Limited Test Ban Treaty, 132
Lindsay Chemical Company, 137
Lippmann, Walter, 75, 83, 87
Lockwood, Stephen T., 20-21
Loeffler, Charles L., 101, 105, 106
Lofftus, James H., 20
Looney, Margaret, 98-100
Los Alamos, New Mexico, 125
Los Angeles County, California, 120
Luminous Processes, Inc., 101

Maggia, Amelia "Mollie," 47-50, 81-88
Maggia, Antonetti, 47
Maggia, Valerio, 47
Maillefer, Sarah Carlough, 68, 75
Manhattan Project, 123-126, 130
Marshall Field Annex building, Chicago, 91
Martland, Harrison S., 4, 66-69, 71-73, 78-79, 82, 87, 109-110, 117, 121-122, 130
Martland, Harrison Stanford Medical Center, 73
Massachusetts Institute of Technology (MIT), 121, 134-135, 136-137
Maximum permissible body burden for radium, 122
Mayo Clinic, 101, 110
McDonald, Quinta Maggia, 47, 77, 79, 81, 88
Mesothorium, 66, 96
Millikan, Robert A., 120-121
Model T, 23, 24
Montclair, New Jersey, 141
Montrose County, Colorado, 18
Moore, Richard B., 87
Morristown, New Jersey, 46
Mount Sinai Hospital, Chicago, 106

Moyar, Charles C., 114, 116
Murphy, Eleanor Flannery, 24

Nagasaki, 126, 130
National Bureau of Standards, 122, 127
National Consumers' League, 60, 61
National Institute for Occupational Safety and Health (NIOSH), 140
National Radium Institute, 27-28
Nazis, 123
Nevada, 131
Nevada Test Site, 131-132
Newark, 42, 48, 55, 67, 72
Newark City Hospital, 72, 73
Newark Public Health Department, 50, 68
New England Watch Company, 136
New Jersey Consumers' League, 60, 80, 82
New Jersey Department of Environmental Protection (DEP), 141-142
New Jersey Department of Labor, 50, 55, 60, 61, 65, 66, 81
New Jersey Radium Research Project, 135
New Jersey State Department of Health, 134
New Jersey workers compensation, 4, 61
New Jersey Zinc Company, 56
New York City, 27, 38, 39, 42, 51, 55, 76, 110, 116
New York Homeopathic Medical College and Hospital, 46
New York Pathological Society, 69
New York Times, 100, 108
New York World, 83
Nobel Prize, 11, 12, 133
    Laureate, 12, 121, 133
    Lecture, Pierre Curie, 13
    Nominating committee in medicine, 72
North Korea, 130
Nuclear testing, aboveground, 6, 131-134

Oak Ridge, Tennessee, 125, 142
Occupational disease, 4, 5, 50
Occupational illnesses, 140
Occupational Safety and Health Administration (OSHA), 139-140
Orange, New Jersey, 1, 5, 42, 47, 58, 60
Orange, New Jersey, *Daily Courier*, 86-87
Orange, New Jersey, health department, 60, 82
Ottawa, Illinois, 5, 91, 99, 100, 104, 105, 108
Ottawa Township High School, 91
Our Lady of Sorrows Roman Catholic Church, Chicago, 107

Pacific proving grounds, 131
Panama Canal, 130
Paradox Valley, Colorado, 18
Paris School of Industrial Physics and Chemistry, 8
Pauling, Linus, 133
Peru, 23
Peru, Illinois, 91, 100
Phillips, Alexander H., 21
Phosphorus poisoning or "phossy jaw," 3, 49
Physicians, 2, 27, 64
Pitchblende ore, 7, 18, 27
Pittsburgh, 22, 24, 26, 28
Placerville, Colorado, 25
Plumb, Robert K., 129
Plutonium, 123-126, 137
Poe, Edgar Allen, 86
Polonium, 7
Pope, Alexander, 121
Poulot, Charles, 19
Princeton University, 21
Proescher, Frederick, 31, 38
Project Sunshine, 133
Prudential Insurance Company, 63
Psychiatric patients, 110, 135
Public health, 6, 87, 142
Public health conference on the nation's radium industries, 87
Purcell, Charlotte, 106

Quack radioactive treatments, current, 141

Radiation sickness, chronic, vii, 5
Radiendocrinator, 112
Radioactive contaminated sites, 141-142
Radioactive fallout, 6, 132-134
Radioactive mill tailings, 141
Radioactivity, 7
Radiograph, 21
Radithor, 112-117
Radium,
    Body burden, 135, 137
    Burns, 12, 45
    Clinics, 38
    Discovery of, 7-8
    Injections, 33, 38-39
    Nasal treatments, 140-141
    Necrosis, 1, 38, 45, 53, 66, 117
    Poisoning, 5, 78-82, 88, 100, 103, 106, 108, 121, 130
    Products, 41-42
    Therapy, 10
    Water, 34-35
Radium Belge, 28
Radium Dial Company, 91-108
    Statement of, 95
Radium Emanator, 35
Radium Institute of New York City, 39
Radium Life Corporation, 35
Radium Luminous Material Corporation, *see* U.S. Radium Corporation
Radium Ore Revigator, 34-35
Radon, 34, 71, 121, 137, 138
    Breath analyzer, 136
    Standard, 122
Rare Metals Reduction Company, *see* Walsh-Lofftus Uranium and Rare Metals Company
Rheumatism, 13, 37, 141
Roach, John, 50-51
Robinson, Mary, 100
Roeder, Arthur, 55-57, 59-60, 65-66, 76, 77
Rollins, William, 31
Roosevelt, Franklin D., 35
Roush, Wade, 139
Rowntree, L. G., 36
Rudolph, Irene, 49-51
Russell Sage Institute, New York, 67

Salida, Colorado, 25
Sarcoma, 5, 100
    Osteogenic, 72
Schaub, Katherine, 77, 79-80, 88
Scientific American, 39, 86
Scintillate, 47
Seaborg, Glenn T., 123-125
Smithsonian Institution, 19
Southampton, Long Island, New York, 117
South Korea, 130
Soviet Union, 130-131
Standard Chemical Company, 24, 26, 27-28
State workers' compensation programs, problems with, 140, 143
Stateville Prison, Joliet, Illinois, 135
Statute of limitations, 81, 89, 102
Staybolt, Tate Flexible, 22, 24
Stevenson, Adlai, 134
Stewart, Ethelbert, 62, 87
St. Joachimsthal, Bohemia, 9, 34, 123
St. Louis Hospital, Paris, 31
Strassmann, Fritz, 123
Strauss, Lewis, 133

Index

Streator, Illinois, 94
Strontium-90, 133, 134
Superfund program, 141
Switzerland, 41, 83
Syphilis, 2, 48, 49, 79, 81

Technological disasters, 142
    Victims of, 142-143
Tetraethyl lead gasoline, 87
Thomas, Norman, 83
Thomas, R. W., 34
Thomson, J. J., 33
Thorium workers, 138
Thorium X, 38
Tiffany & Company, 41
Time magazine, 85
Truman, Harry S., 123, 126
Tympanotomy tubes, 141

Ukraine, 42
University of California, Berkeley, 123, 125
University of Chicago, 68, 138
University of Illinois, 105, 106
University of Lvov, 42
University of Moscow, 42
University of Paris, 11, 12
University of Pennsylvania, 15, 51
University of Vienna, 51, 110
Uranium, 7, 20
Uranium-235, vii, 123
U.S. Atomic Energy Commission (AEC), vii, 6, 127, 130, 132-133
U.S. Bureau of Labor Statistics, 62, 65
U.S. Bureau of Mines, 20, 27-28, 87
U.S. Department of Energy, 6, 142
U.S. Department of Labor, 96, 99
U.S. Environmental Protection Agency, 141-142
U.S. Food and Drug Administration (FDA), 117, 121

U.S. Navy, 119
    Medical Corps, 122
U.S. Public Health Service, 61, 69, 87
U.S. Radium Corporation, 4, 29, 41, 45, 55, 59-60, 64, 66, 68-71, 75-77, 79-80, 82-83, 85, 89, 113, 135
U.S. Supreme Court, 108
Utah, 20-21, 24, 27, 132, 134

Vallat, Inez Corcoran, 101-102
Vanadium, 18, 21, 23, 24
Vichy, France, 34
Voilleque, Charles, 19
Von Sochocky, Sabin Arnold, 42-43, 45-46, 82

Walsh, John, 20
Walsh-Lofftus Uranium and Rare Metals Company, 21
Waterbury Clock Company, 44, 71, 135, 136, 142
Waterbury, Connecticut, 5, 44, 71, 136
WCFL Radio Station, Chicago, 108
West Berlin, 130
West Chicago, Illinois, 137
Westclox, 91, 94
Western Clock Manufacturing Company, *see* Westclox
West Orange, New Jersey, 17, 135
Wiley, Katherine G. T., 60-65, 82
Willis, George S., 42, 46
Workplace hazards, 140-141
Work-related diseases, 140
World War I, 1, 42-43, 119
World War II, 6, 118, 122, 126, 129, 140

X-rays or Roentgen rays, 7, 15-16, 31, 58, 86, 105, 109, 116, 121

Yale University, 85, 114